INTEGRATING LECTURE AND LAB

A GENERAL BIOLOGY LABORATORY MANUAL

Bassim Hamadeh, CEO and Publisher
Carrie Montoya, Manager, Author Care and Revisions
Kaela Martin, Project Editor
Alia Bales, Production Editor
Jess Estrella, Senior Graphic Designer
Alexa Lucido, Senior Licensing Specialist
Natalie Piccotti, Director of Marketing
Kassie Graves, Vice President of Editorial
Jamie Giganti, Director of Academic Publishing

Printed in the United States of America.

ISBN: 978-1-5165-1792-3 (pbk) / 978-1-5165-1793-0 (br)

cognella | ACADEMIC PUBLISHING

INTEGRATING LECTURE AND LAB

A GENERAL BIOLOGY
LABORATORY MANUAL

Third Edition

By Leslie A. King

University of San Francisco

❀ cognella® | ACADEMIC PUBLISHING

CONTENTS

PREFACE AND INTRODUCTION TO TAXONOMY

The activities in this manual are designed to complement and reinforce the material you will learn in the lecture portion of this course. The first half of the labs provides an overview of living organisms and their taxonomy, and the remaining exercises focus on the comparative physiology of many of the animal groups studied in the first half.

In the 1700s, Linnaeus devised the hierarchical system of classification we still use today. In the simplest version of this system, the taxonomic categories are, from the broadest to the most specific:

Domain
Kingdom
Phylum
Class
Order
Family
Genus
Species

Other categories such as "subphylum" or "superorder" are sometimes used as well. The genus and species of an organism comprise its scientific name; for example, *Homo sapiens* is the scientific name for a human and *Carcharodon carcharias* is the scientific name for the great white shark. The genus name comes first and is capitalized, and both names are in Latin and are italicized. To note here is the difference between the terms taxonomic category and taxon (pl. = taxa). Taxonomic categories are the list above (domain through species). A taxon is a specific category name. For example, sharks are in the class Chondrichthyes; "class" is the taxonomic category, and "Chondrichthyes" is the taxon.

As will be mentioned throughout the semester, taxonomy is dynamic. The classification of some groups, such as the protists, has been a source of debate among taxonomists for years. As scientists began using molecular techniques to investigate the relatedness of organisms, new possible evolutionary relationships began to emerge: for example, organisms previously thought of as being somewhat unrelated (or distantly related) were shown to have very similar DNA sequences for the genes studied. As a result, phylogenetic trees were rearranged and they continue to be reconstructed as more data are collected. The taxonomy of organisms is an ongoing project, and it is possible to see different classifications of the same organism in different sources. For lab this semester, you should refer to the classification discussed in lecture and in your textbook.

Because of the dynamic nature of taxonomy, especially in certain groups, this manual presents taxon names only for those organisms/groups whose classification remains somewhat consistent from one source to another. For other organisms, the common names are provided and you should consult your textbook and lecture notes for the formal taxa to learn.

GENERAL LABORATORY SAFETY GUIDELINES

You must assume responsibility for the safety of yourself and others working in the lab. The following are some safety and procedural rules to help guide you in protecting yourself and others from injury in the laboratory. These rules must be followed at all times.

1. Conduct yourself in a responsible manner at all times in the laboratory.
2. Do not eat food, drink beverages, smoke, or chew gum in the laboratory.
3. Work areas should be kept clean and tidy at all times. Only have required materials out on your lab bench. Other materials (e.g., purses, backpacks, coats, and cell phones) should be stored out of the way in the space provided.
4. Know the locations and operating procedures of all safety equipment, including the first aid kit, eyewash station, safety shower, spill kit, and fire extinguisher. Familiarize yourself with the locations of the fire alarm and the room and building exits.
5. Know what to do if there is a fire drill: Any open gas valves should be turned off and any electrical equipment should be turned off. Follow the directions provided by your lab instructor.
6. Perform only those experiments and use only equipment for which you have received training. Carefully follow all instructions, both written and oral.
7. Do not wear open-toed shoes; feet should be completely covered when you are in the lab. Gloves should be worn whenever you use hazardous chemicals. DO NOT use your cell phone or calculator with gloves on, as this could pass on potentially harmful organisms and chemicals that you have handled to your phone and eventually to you.
8. Dispose of all chemical and biological waste properly in labeled containers. Pay attention to directions given by your lab instructor.
9. If at any time you are not sure how to handle a particular situation, ask your lab instructor for advice. It is always better to ask questions than to risk harm to yourself or damage to the equipment.
10. Report all accidents, no matter how minor, to your laboratory instructor immediately.

PROKARYOTES

INTRODUCTION

Members of the two prokaryotic domains, Bacteria and Archaea, are ubiquitous, inhabiting every terrestrial and aquatic habitat on the planet. In addition, prokaryotes (along with a variety of fungi and viruses) are important members of the human microbiome, living on and within the human body. In fact, the presence of certain bacterial species in the gastrointestinal tract is essential to digestive health in people. On the other hand, some bacteria are pathogenic, causing disease in humans and other organisms.

A large part of the success of the prokaryotes is due to their metabolic diversity. They are diverse in both how they produce ATP (phototrophic or chemotrophic) and in how they acquire organic molecules (autotrophic or heterotrophic.) Some bacteria carry out nitrogen fixation, a process in which atmospheric nitrogen (N_2) is converted to ammonia (NH_3). This is extremely important ecologically, since most plants are unable to use N_2 ; the NH_3 produced via nitrogen fixation then gets converted by other bacteria into forms of nitrogen that are usable by plants.

ACTIVITIES

Exercise 1. Investigation of Bacterial Colonies and Cell Morphology
NOTE: Gloves must be worn for any activity involving the handling of bacterial cultures.

A. Observation of bacterial colony morphology
A bacterial colony is a macroscopic cluster of cells that originated from (usually) a single cell. Therefore, the colony essentially is a clone of cells. Observation of colony morphology, color, and texture can be useful in initially identifying a bacterial species. (Figure 1).

Whole colony:

Punctiform

Circular

Rhizoid

Irregular

Filamentous

Edge:

Entire

Undulate

Lobate

Filamentous

Curled

Surface:
Smooth, glistening
Rough
Wrinkled
Dry, powdery

Elevation:
Flat
Raised
Convex
Pulvinate
Umbonate

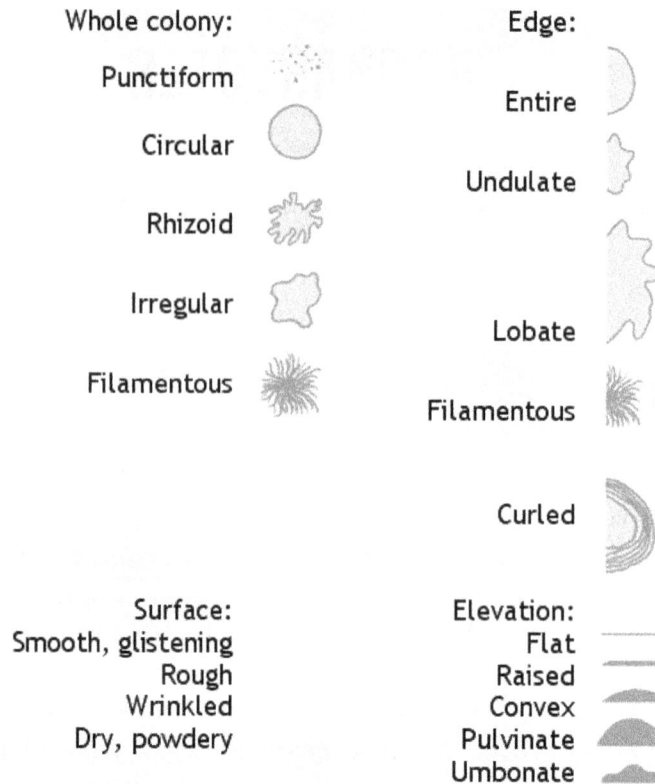

FIGURE 1. **Bacterial colony morphologies**

PROCEDURE

1. Observe any bacterial cultures on display and note any differences in colony shape, texture, size, and color.

B. Observation of individual cell morphology

Most bacterial cells exhibit one of three main shapes: 1) **bacilli** (rod-shaped cells; singular: bacillus); 2) **cocci** (spherical cells; singular: coccus); or 3) **spiral-shaped**. Depending on the particular species, bacterial cells may exist individually or may be arranged in grape-like clusters, in chains, or in pairs. Species with their cells arranged in clusters are described using the prefix "staphylo-" in their name or description, e.g., *Staphylococcus aureus*. Species with cells arranged in chains have the prefix "strepto-" in their name or description; e.g., *Streptococcus pneumoniae*. Species with cells arranged in pairs have the prefix "diplo-" in their name or description, for example, diplobacillus.

PROCEDURE

1. Working individually, observe prepared slides of bacilli, cocci, and spiral-shaped bacteria using the compound microscope. Remember to begin your focusing by using the 4x or 10x objective.

Once the cells are in focus at these magnifications, you may switch the objective to 40x and use the fine focus knob to bring the cells into focus.

2. Once the cells are in focus under the 40x objective, view the bacteria using the oil immersion (100x) objective lens. To do this, rotate the 40x objective out of the way and put a drop of oil directly on the area of the slide you are viewing and rotate the 100x objective into place. Note that you do NOT need to move the stage before moving the 100x into position; the 100x lens is supposed to touch the oil. When you are finished viewing the specimen using the 100x lens, rotate the 4x objective (NOT the 40x objective) back into position. This will ensure that no oil gets on the 40x objective.

3. Using lens paper, wipe the oil from the 100x objective lens and also wipe the oil from the slides.

C. Observation of dental plaque

Dental plaque consists of a film of bacteria that grows on the teeth.

PROCEDURE

1. Observe a prepared slide of a dental plaque smear. The slide has been Gram-stained for visualization. In addition to bacterial cells, this smear likely also contains human epithelial cells obtained while preparing the slide. Note the difference in size between the epithelial cells (=eukaryotic) and bacterial cells.

Exercise 2. The Gram Stain

The Gram stain is a differential staining technique named after its inventor, the Danish microbiologist Hans Christian Gram. It is used to classify a bacterial species into one of two large groups, Gram-positive bacteria or Gram-negative bacteria. These differ in their cell wall structure and therefore in their ability to retain crystal violet, a purple stain.

The cell walls of all bacteria contain peptidoglycan, a polysaccharide that is cross-linked by short amino acid chains. Gram-positive bacterial cell walls possess a thick layer of peptidoglycan, while Gram-negative bacterial cell walls consist of a thinner layer of peptidoglycan surrounded by an outer phospholipid bilayer. Gram-positive cells retain crystal violet and stain purple. Gram-negative cells do not retain the stain; they are pink after the staining procedure. The pink color results from the use of safranin, a counterstain (Figure 2.)

Other techniques exist for distinguishing between Gram-positive and Gram-negative bacteria in a sample. Determining whether a bacterial species is Gram-positive or Gram-negative can be an important preliminary step in the detection of and treatment of a suspected bacterial infection.

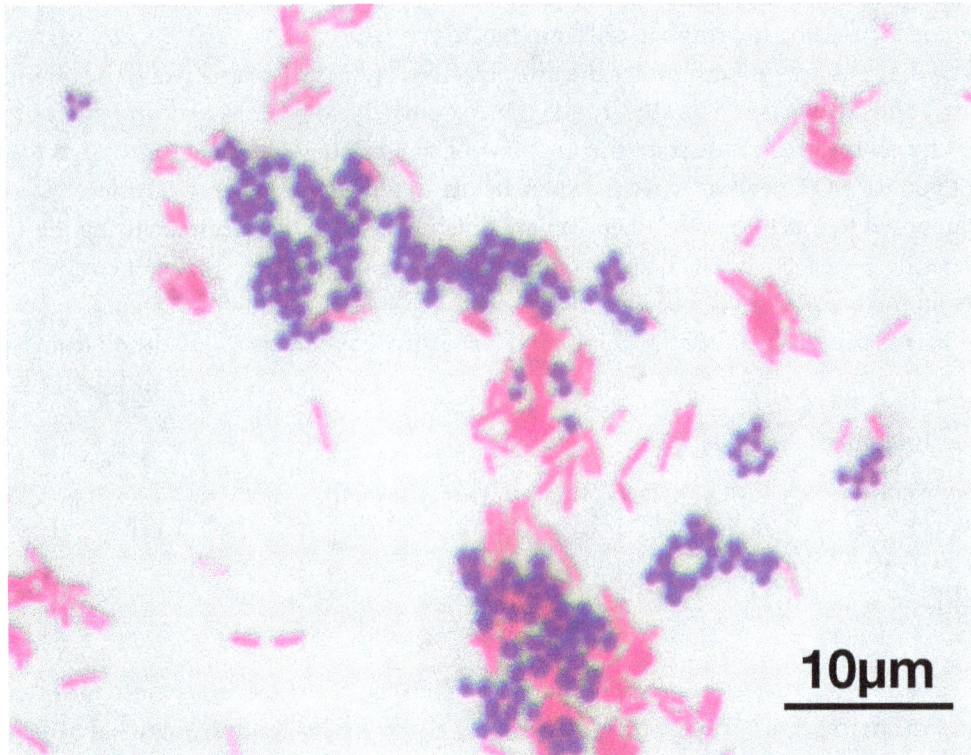

FIGURE 2. Gram stain showing Gram-positive cocci *(Staphylococcus aureus)* **and Gram-negative bacilli** *(Escherichia coli).*

TABLE 1. Summary of Chemicals Used in Gram Staining.

Chemical	Function
crystal violet (a purple dye)	crosses bacterial cell wall and binds to cell membrane, making the cells purple in color
iodine solution (Gram's iodine or I_2KI)	a mordant: fixes the crystal violet, makes it insoluble
95% ethanol	a decolorizing agent: removes excess stain from cells, removes the purple dye complex from Gram (-) bacteria: dissolves outer lipid layer on Gram (-) cells; Gram (+) cells retain the stain
safranin	a counterstain: stains Gram (-) bacteria pink so that they may be visualized (Gram (+) cells' purple color masks out the pink of safranin)

A. Observation of Gram-positive and Gram-negative bacteria

PROCEDURE

1. Obtain prepared slides of Gram-positive and Gram-negative bacteria and observe under the compound microscope.
2. Locate the liquid bacterial cultures provided in lab. Note the bacterial species growing in each culture:

Species A: _____

Species B: _____

Perform a Gram stain or alternate technique to determine whether these bacterial species are Gram-positive or Gram-negative. Further instructions will be given in lab.

Exercise 3. Investigation of Bacteria in the Environment

As previously mentioned, bacteria are ubiquitous. They are found in nearly every environment on earth, including on and within living organisms, in soil, in and on food, and on many surfaces with which humans come in contact. In this exercise, you will investigate the presence of bacteria in an environment of your choice.

PROCEDURE

1. Form a group with other lab members and decide upon an environment to investigate. Examples of possible environments are: a door knob, lab bench, drinking fountain handle, backpack, refrigerator drawer, etc.

Environment: _____

Talk to other groups in your lab. What are some other environments being investigated?

2. Before you begin the experiment, hypothesize about the presence of bacteria in your environment. Do you expect your environment to contain a greater number of bacterial cells than other environments being tested? Do you think your environment contains a wide variety of bacterial species, or is it limited to perhaps one or two species? (Note that you will not be able to determine the exact species found in your environment, or the exact number of bacterial cells, but in the next lab, you will be able to count colony numbers in your culture and will be able to observe similar or different colony types.) Record your hypotheses below:

Hypotheses regarding bacteria living in the environment of choice: _____

3. Obtain a TSA (tryptic soy agar) plate and label the bottom of the plate with your name (or group members' initials), date, lab section number, and your environment name.
4. Using a sterile swab, immediately take <u>one</u> swab from your assigned environment. Do not swab repeatedly. Keep track of which side of the swab contacted the surface.
5. Without touching the swab to any surface, remove the lid from your TSA plate. Using the isolation streak technique described by your instructor, swab the plate with the same side of the swab that originally contacted your environment. Streaking the plate in this way allows for isolation of individual bacterial strains and should, after incubation, allow you to view individual colonies of bacteria rather than a lawn of bacterial growth.
6. Replace the lid on the plate and put the swab in the biohazard bag.
7. Place your plate in the area designated for transfer to the incubator.

RESULTS

1. During the next laboratory period observe your plates and record your observations in Table 2, sharing your results with the class to fill in the rest of the table. For the "colony diversity" column, count the number of different colony types you see that differ in color, shape, size, and texture. Note that colonies of different bacterial species can resemble each other, so this may not be a complete count of different species that grew on the plate.

Exercise 4. Investigation of the Effects of Antibiotics on Bacterial Growth

Antibiotics are chemicals produced by microorganisms that inhibit or slow the growth of other micro-organisms. Antibiotics inhibit bacterial growth in several ways. For example, some antibiotics work by inhibiting cross-linking of peptidoglycan in cell wall formation, while others work by inhibiting a certain step in protein synthesis. Some antibiotics, because of their mechanism of action, are effective against only certain families of bacteria, only Gram-positive bacteria, or only Gram-negative bacteria. Other antibiotics are effective against both Gram-positive and Gram-negative bacteria and are called broad-spectrum antibiotics.

In today's lab, you will expose a bacterial species to different antibiotics; and then next week, you will observe which antibiotics were more effective at preventing the growth of the bacteria. To expose the bacteria to the antibiotics, you will place five disks, each impregnated with a different antibiotic, onto an agar plate that has been swabbed with a bacterial culture. The antibiotics in the disks will diffuse into the agar. If a particular antibiotic is effective at preventing the growth of your bacterial species, you will see a "halo" of no growth, called a zone of inhibition, around the disk (Figure 3).

TABLE 2. Results of Incubation of TSA Plates Containing Bacteria Swabbed from Different Environments. (Plates were incubated for 24–48 hours at 37°C.)

Environment	Colony Diversity (# of different colony types visible)	Colony Color(s)	Amount of Bacterial Growth (sparse, moderate, or abundant)?

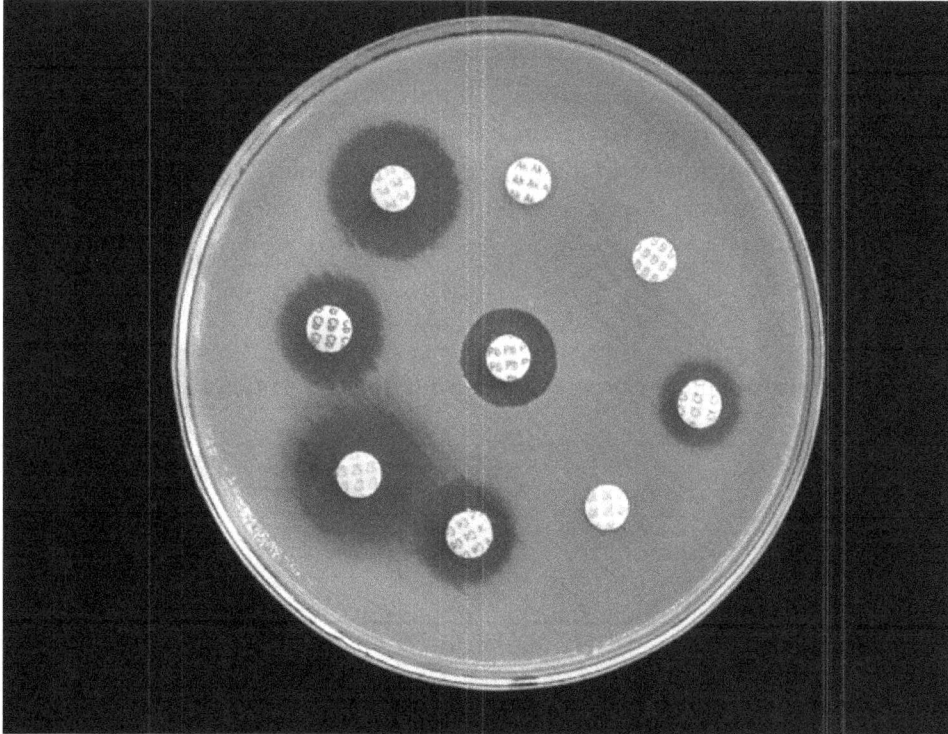

FIGURE 3. Antibiotic susceptibility testing using disks impregnated with different antibiotics. "Halos" of no growth indicate inhibition of growth by the antibiotic.

PROCEDURE

Lab 1 (today) (NOTE: Work in the same groups you were in for Exercise 3)

1. Obtain a TSA plate and label the bottom of the plate with your name (or group members' initials), date, lab section number, and your assigned bacterial species.
2. Soak a sterile cotton swab in a liquid culture of your assigned bacteria, remove the lid to your plate, and carefully swab the entire agar plate to ensure complete coverage over the entire surface of the plate. Note that this is not the isolation streak technique you used in Exercise 3; for this exercise, the aim is to ensure a lawn of growth on the plate, not individual colonies.
3. Using clean forceps, firmly place the four different antimicrobial disks (symbol side up) on your plate, making sure to evenly distribute them. Your lab instructor may choose to use different antibiotics than those listed.
4. Replace the lid on the plate and put the swab in the biohazard bag.
5. Place your plate in the area designated for transfer to the incubator.

Lab 2 (next lab period)

1. Measure and record the zone of inhibition in millimeters in Table 3 or Table 4 (depending on your bacterial species) to determine the bacterial sensitivity for each antibiotic. The zone of inhibition is the distance between the edge of the antimicrobial disk and the edge of the "halo" of no growth. If there was no zone of inhibition, record the size as 0.
2. Exchange results with a group that worked with the other bacterial species and record in the corresponding table.

TABLE 3. Results of Exposure of *Staphylococcus epidermidis* to Four Antibiotics. (Plates were incubated for 24–48 hours at 37°C.)

Antimicrobial Agent	Zone Size (mm)
Vancomycin	
Erythromycin	
Penicillin	
Tetracycline	

TABLE 4. Results of Exposure of *Escherichia coli* to Four Antibiotics. (Plates were incubated for 24–48 hours at 37°C.)

Antimicrobial Agent	Zone Size (mm)
Vancomycin	
Erythromycin	
Penicillin	
Tetracycline	

Exercise 5. Observation of Cyanobacteria

Cyanobacteria are aquatic, photosynthetic prokaryotes. In fact, data strongly support the cyanobacteria as being the first organisms to put oxygen into the atmosphere. A variety of pigments can be found in different cyanobacterial species, but they get their name from the presence of a blue pigment called phycocyanin. Cyanobacteria also possess the green pigment chlorophyll a. Due to the bluish-green color of many species, they are sometimes called "blue-green algae," but this name is misleading since algae are eukaryotic. As mentioned previously, evidence shows that the first organisms to put oxygen into the atmosphere (via photosynthesis) were cyanobacteria. In addition to being photosynthetic, many cyanobacteria carry out nitrogen fixation.

Refer to your lecture notes and textbook for more information on cyanobacteria.

PROCEDURE: MAKING A WET MOUNT

To observe live microscopic organisms in today's lab and in future labs, you will need to prepare wet mounts of the organisms.

To make a wet mount from a culture of live organisms:

1. Withdraw a sample from the culture using a transfer pipette.
2. Place a drop of the culture on a clean glass slide and gently drop a coverslip at an angle onto the sample.

A. *Anabaena* sp.

Anabaena is a freshwater cyanobacterium that grows in long filamentous chains (Figure 4). *Anabaena* possesses structures called heterocysts at intervals along the filament. Heterocysts carry out nitrogen fixation, providing neighboring cells with nitrogenous compounds.

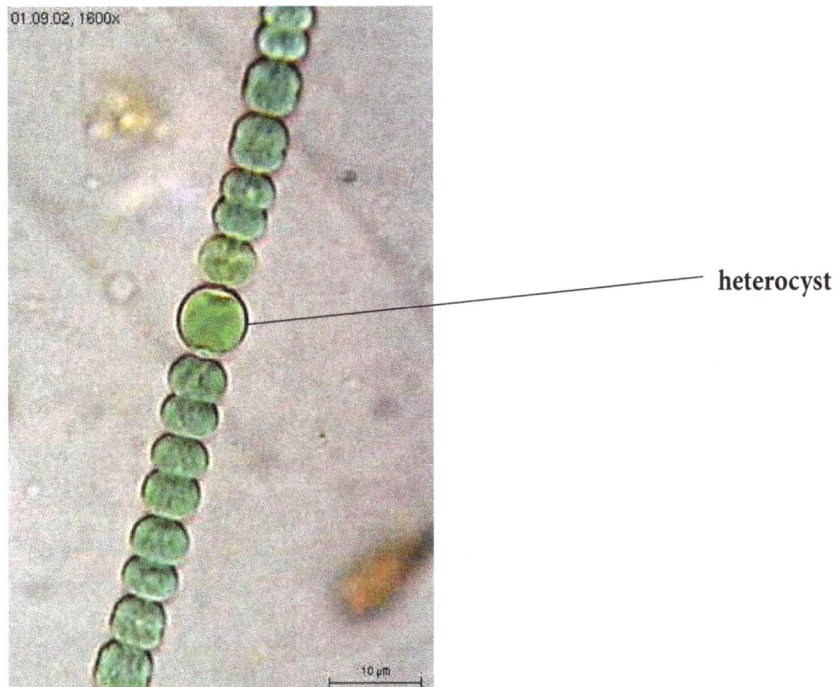

FIGURE 4. *Anabaena* sp. with heterocyst.

1. Prepare a wet mount from a live culture of *Anabaena* and observe using the compound microscope. Draw a representative *Anabaena*. (Note: The culture you are viewing may contain other live organisms in addition to *Anabaena*.) Label any heterocysts that you see.

B. *Oscillatoria* sp.

Oscillatoria is another filamentous, freshwater cyanobacterium (Figure 5). This organism gets its name from the oscillating movement it exhibits.

FIGURE 5. *Oscillatoria* sp.

1. Prepare a wet mount from a live culture of *Oscillatoria* and observe, using the compound microscope. Draw a representative *Oscillatoria*. (Note: The culture you are viewing may contain other live organisms in addition to *Oscillatoria*.)

C. *Spirulina* sp.

Spirulina is a spiral-shaped cyanobacterium found in both marine and freshwater environments (Figure 6).

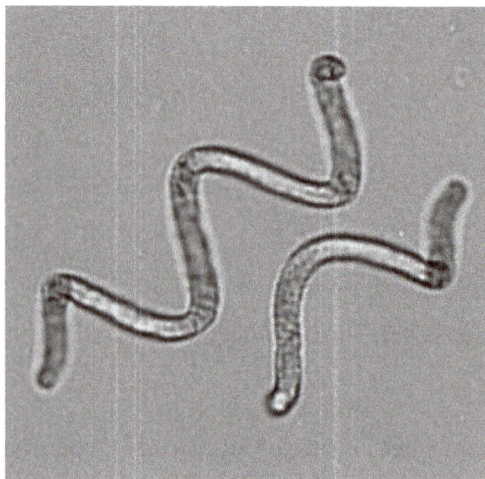

FIGURE 6. *Spirulina* sp.

1. Prepare a wet mount from a live culture of *Spirulina* and observe, using the compound microscope. Draw a representative *Spirulina*. (Note: The culture you are viewing may contain other live organisms in addition to *Spirulina*.)

QUESTIONS

1. What are some main characteristics of prokaryotes? How do prokaryotes and eukaryotes differ?

2. How do the cell walls of Gram-positive and Gram-negative bacteria differ from one another?

3. What is nitrogen fixation? What is its importance?

4. What are heterocysts? In what bacterial group are heterocysts found?

5. Describe ways in which bacteria (including the cyanobacteria) are important ecologically. How are the cyanobacteria important from an evolutionary standpoint?

QUESTIONS TO BE COMPLETED IN THE NEXT LAB PERIOD:

6. Discuss the results of Exercise 3. How did the bacterial growth on your plate compare with your hypothesis? How did the bacterial growth on your plate compare with the other environments investigated? Which environment(s) in your class appeared to have the most growth or greatest abundance of bacterial colonies? The greatest colony diversity?

7. According to the results of Exercise 4, which antibiotics were most effective against each organism? Did the results differ for Gram + and Gram - bacteria? Explain.

FIGURE CREDITS

Figure 1: Source: http://commons.wikimedia.org/wiki/File:Bacterial_colony_morphology.png.

Figure 2: Copyright © Y tambe (CC BY-SA 3.0) at http://commons.wikimedia.org/wiki/File:Gram_stain_01.jpg.

Figure 3: Source: http://commons.wikimedia.org/wiki/File:Antibiotic_disk_diffusion.jpg.

Figure 4: Copyright © קאלדנב הנוי (CC BY-SA 3.0) at http://commons.wikimedia.org/wiki/File:Anabaena_sperica2.jpg.

Figure 5: Copyright © NEON (CC BY-SA 2.5) at http://commons.wikimedia.org/wiki/File:Oscillatoria_sp.jpg.

Figure 6: Copyright © Joan Simon (CC BY-SA 3.0) at http://commons.wikimedia.org/wiki/File:Spirul2.jpg.

PROTISTS AND AN INTRODUCTION TO PLANTS

INTRODUCTION

The protists are an extremely diverse group of eukaryotes. Any eukaryote that is not a fungus, an animal, or a land plant is a protist. Many protists are unicellular, and in those that are multicellular, there is little differentiation of cells into tissues. Protists vary greatly in their morphology, in their modes of nutrition and reproduction, and in their mechanism of locomotion, and they are found in a variety of habitats. Many photosynthetic protist species are important ecologically in that they are primary producers in aquatic habitats. Some protists affect human health by causing diseases such as malaria, sleeping sickness, and giardiasis, to name a few.

The classification of protists has been the subject of strong debate among taxonomists for years. In this lab, you will be observing organisms from a variety of protist lineages and you will be asked to identify each. As mentioned in the preface, the taxonomy of protists (and of all organisms) is ongoing and no doubt you will see different classifications in different textbooks. For this lab, you should refer to the classification discussed in lecture and in your textbook.

The green algae, along with the land plants, are often grouped into the "green plants." Although green algae are still referred to as protists in many sources, they are also considered to be green plants because of their many similarities with land plants. In fact, data suggest that land plants arose from a multicellular freshwater green algal ancestor. Green algae and land plants both have chlorophylls a and b, both have cellulose in their cell walls and possess plasmodesmata, and both groups synthesize starch as a storage carbohydrate. Today you will observe examples of unicellular, colonial, and multicellular green algae.

ACTIVITIES

Exercise 1.

Observe the bacterial cultures you prepared for Exercises 3 and 4 in last week's lab. Record your observations and results in Tables 2, 3, and 4.

Exercise 2. Review of Alternation of Generations

Several protist groups and all of the land plants exhibit alternation of generations, a life cycle characterized by a sporophyte phase and a gametophyte phase. The diploid sporophyte produces haploid spores via meiosis in structures called sporangia. The spores are released from a sporangium, and each is capable of developing into a new haploid gametophyte. The gametophyte produces haploid gametes via mitosis in structures called gametangia. Antheridia are gametangia that produce sperm, while archegonia are gametangia that produce eggs. More details of alternation of generations will be presented as the life cycles of different land plants are studied.

Using your lecture notes, lab notes, and your textbook, draw a general diagram of alternation of generations in the space below. In your diagram, include the following terms: gametophyte, spore, gametes, mitosis, meiosis, and fertilization (sporophyte is given to help you begin). Also indicate the ploidy level (1n or 2n) of the following: sporophyte, gametophyte, spore, gametes.

Exercise 3. Observation of Representative Protists

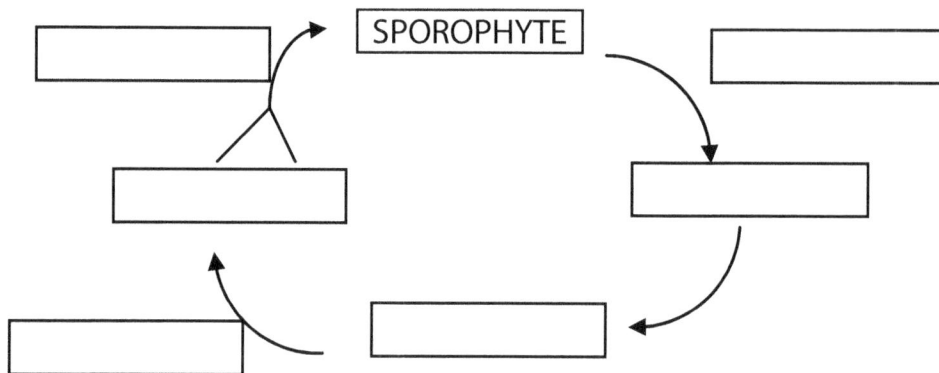

A. *Trypanosoma brucei*

Trypanosoma brucei is a unicellular parasite that causes African sleeping sickness in humans and other mammals (Roberts and Janovy, 2009). Structural features of a trypanosome that are visible under a compound microscope are a flagellum, nucleus, an undulating membrane, and a kinetoplast (a mass of mitochondrial DNA) (Figure 7).

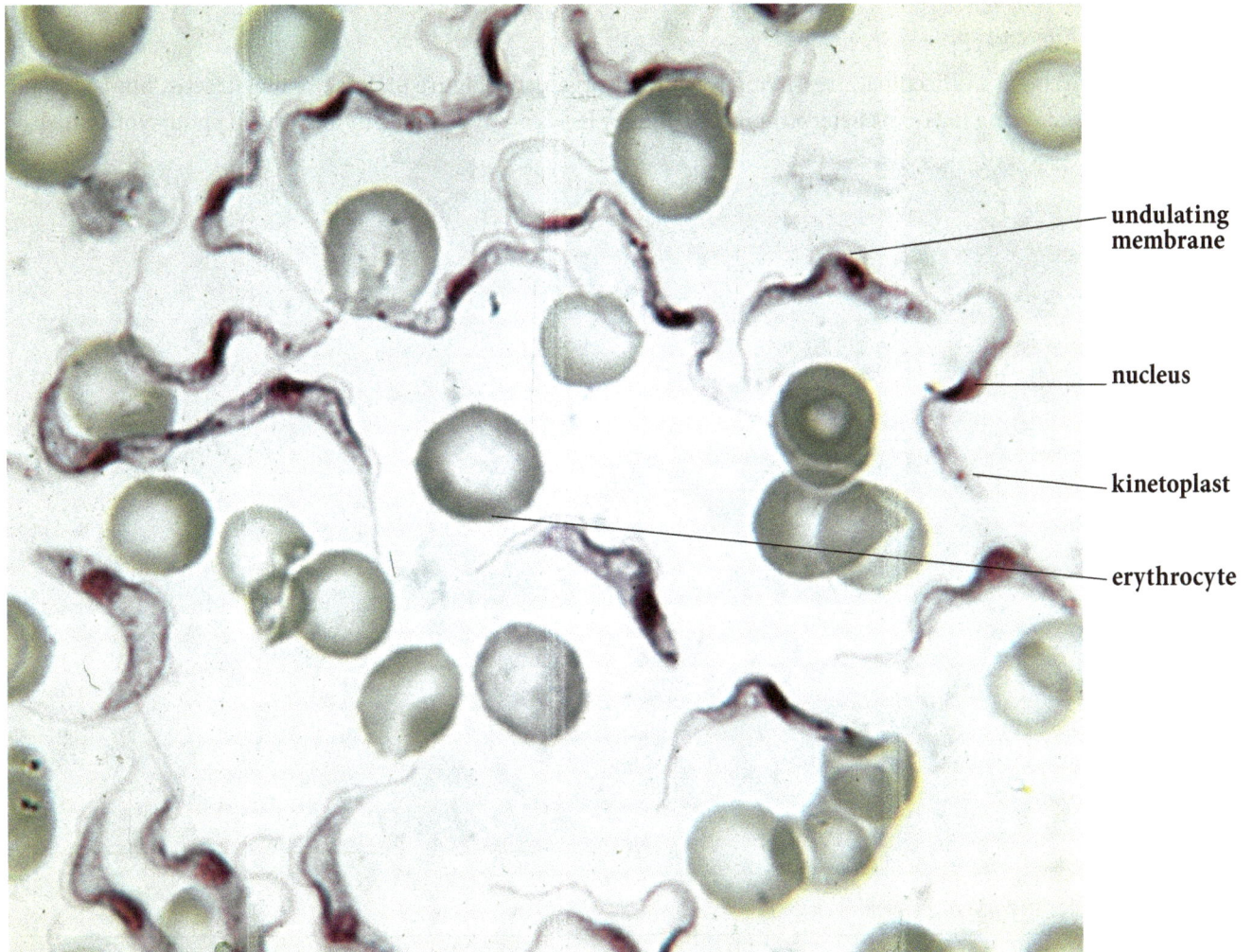

FIGURE 7. *Trypanosoma brucei.*

1. Observe a prepared slide of *Trypanosoma* using the compound microscope. Draw a representative trypanosome, viewed using the 40x or oil immersion (100x) lens. These slides are blood smears that were made with the blood of infected rats, so you will also see erythrocytes, leukocytes, and platelets in the sample. Label the parasite's nucleus, flagellum, undulating membrane, and kinetoplast. Also draw a red blood cell, which typically measures 7 micrometers in diameter. Using the diameter of a red blood cell as a reference, how long is *Trypanosoma*?

2. Turn to Table 5 at the end of this chapter. Fill in the parts of the chart that correspond to *Trypanosoma* (Hint: The first step in filling out the chart is placing *Trypanosoma* into the "example" square that corresponds to its lineage and group).

B. *Euglena* sp.

Euglena is a unicellular, free-living freshwater organism (Figure 8). It is photosynthetic, but is capable of obtaining nutrients heterotrophically if light levels are not sufficient to support photosynthesis.

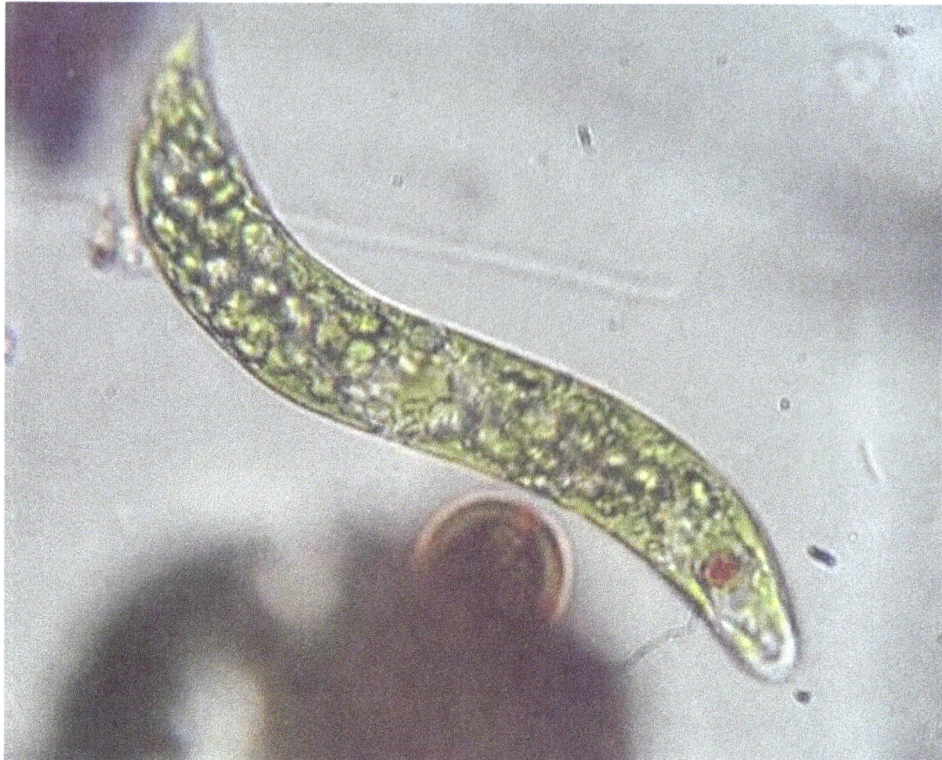

FIGURE 8. *Euglena* sp.

1. Observe a prepared slide of *Euglena* using the compound microscope. Note the following structures: flagellum, eyespot, and chloroplast. (Note: The flagellum is not visible in Figure 8.)
2. Prepare a wet mount from a live culture of *Euglena* and observe using the compound microscope. (If you need a reminder of how to prepare a wet mount, turn to the cyanobacteria portion of the prokaryotes lab.) For best viewing of wet mounts, close the condenser's iris diaphragm, using the lever on the front of the iris to enhance contrast of the specimen.
3. Draw a representative *Euglena*. Indicate the objective lens and total magnification used to view the specimen. (Note: The culture you are viewing may contain other live organisms in addition to *Euglena*.) Label the flagellum and chloroplast.
4. Turn to Table 5 at the end of this chapter and fill in the parts of the chart that correspond to *Euglena*.

C. *Paramecium caudatum*

Paramecium caudatum is a unicellular, free-living freshwater heterotroph (Figure 9). Like all ciliates, *Paramecium* moves via rows of cilia, possesses a macronucleus and micronucleus, expels water using a contractile vacuole, and possesses an oral groove that it uses to feed.

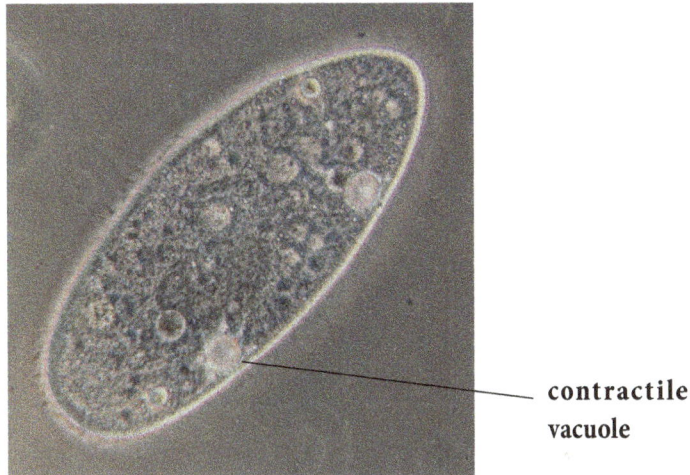

contractile vacuole

FIGURE 9. *Paramecium* **sp.**

1. Observe the *Paramecium* model and identify the following: macronucleus, micronucleus, contractile vacuole, food vacuoles, oral groove.
2. Observe a prepared slide of *Paramecium* using the compound microscope. Note the cilia, oral groove, and nucleus (it may not be possible to distinguish between the macro- and micronucleus).
3. Prepare a wet mount from a live culture of *Paramecium* and observe, using the compound microscope.
4. Draw a representative *Paramecium*. Indicate the objective lens and total magnification used to view the specimen. (Note: The culture you are viewing may contain other live organisms in addition to *Paramecium*.) Label the cilia, oral groove, contractile vacuole, and nucleus (as with the prepared slide, it may not be possible to distinguish between the macro- and micronucleus).
5. Turn to Table 5 at the end of this chapter and fill in the parts of the chart that correspond to *Paramecium*.

D. *Plasmodium* sp.

Plasmodium sp. is an intracellular parasite of red blood cells that causes malaria (Figure 10). Refer to your lecture and lab notes along with your text for more information on *Plasmodium* sp.

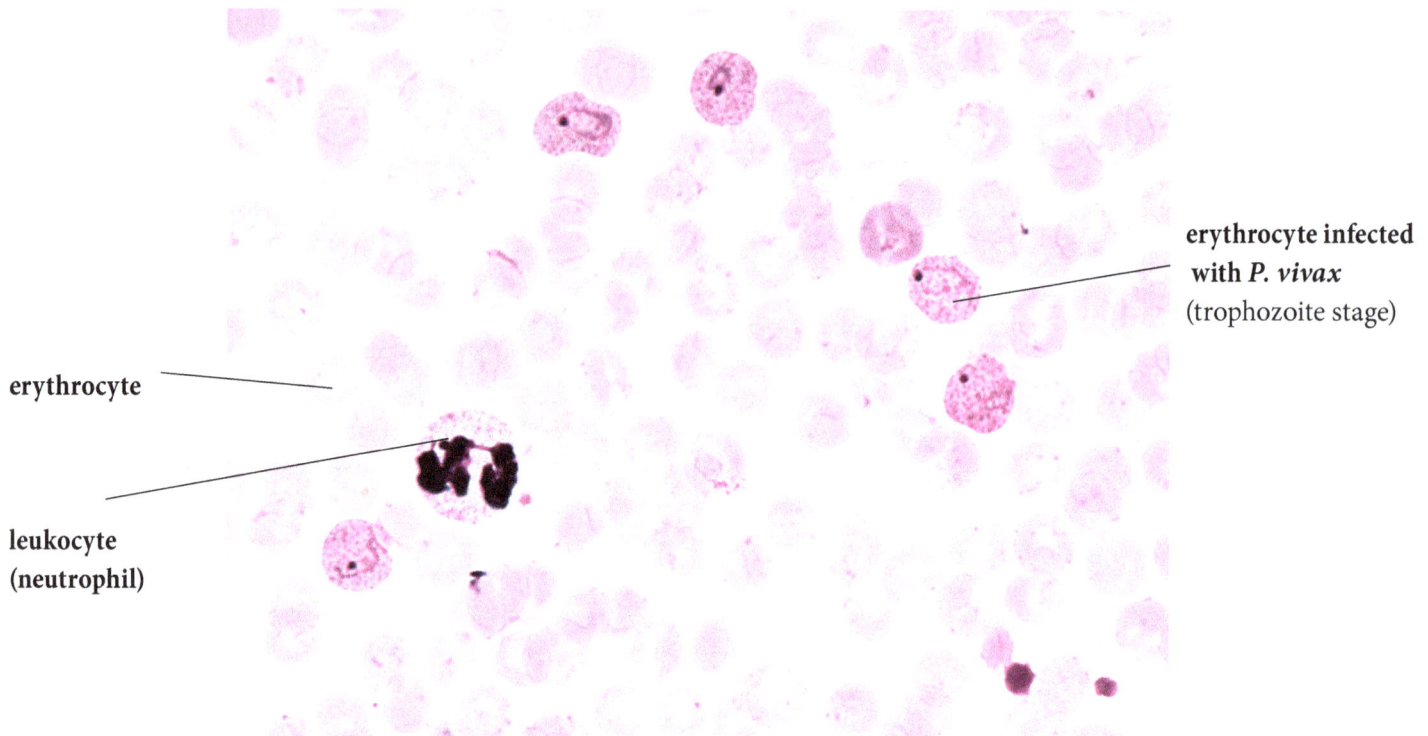

erythrocyte infected
with *P. vivax*
(trophozoite stage)

erythrocyte

leukocyte
(neutrophil)

FIGURE 10. *Plasmodium vivax* in human blood, 1142x.

1. Observe the prepared slide of red blood cells infected with *Plasmodium* on demonstration and draw what you see.
2. Turn to Table 5 at the end of this chapter and fill in the parts of the chart that correspond to *Plasmodium*.

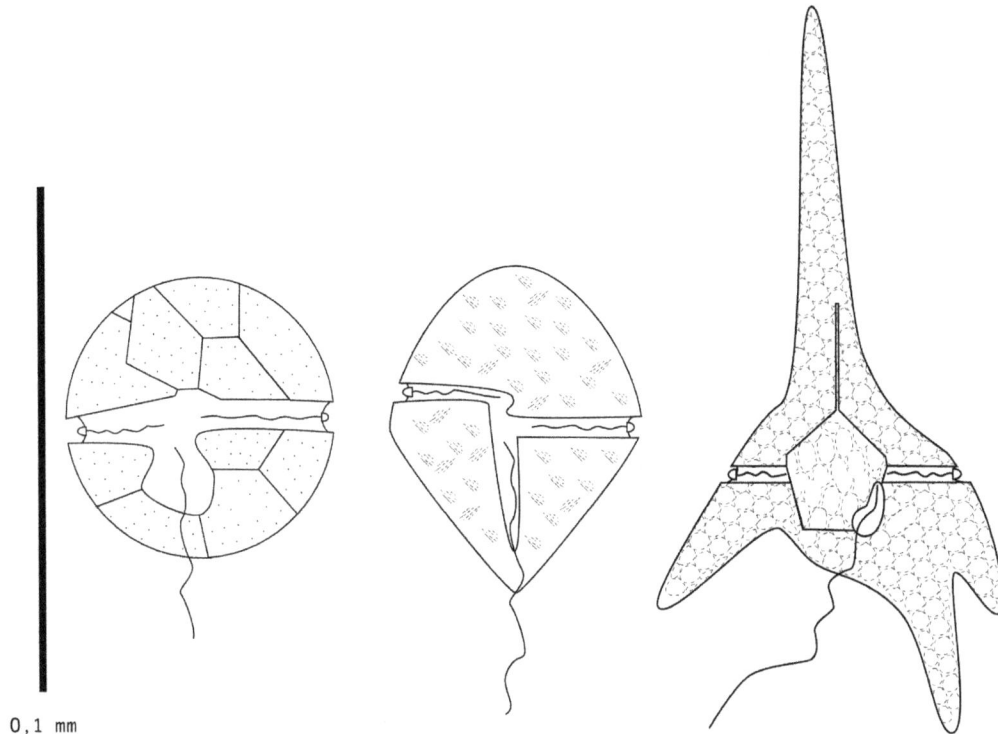

0,1 mm

FIGURE 11. **Representative dinoflagellates.**

E. Dinoflagellates

Dinoflagellates are unicellular photosynthetic organisms. There are both marine and freshwater species, and they play important roles in aquatic food chains. Most are free-living, but some dinoflagellates form symbiotic relationships with some animals (for example, corals and nudibranchs) by living in the tissues of these organisms and providing food via their photosynthetic activity. Some species of dinoflagellates are toxic, and when they experience rapid population growth (a "bloom"), the toxins produced can accumulate in the tissues of filter-feeding organisms such as clams and mussels. Usually these organisms are not harmed by the high levels of toxins in their tissues, but humans who eat contaminated animals can be poisoned. A bloom of toxic dinoflagellates in the water is often called a red tide because of the coloration of the water by the red or reddish-brown pigments found in many toxic species.

1. Observe a prepared slide of dinoflagellates using the compound microscope. In order to see detail, you will need to use either the 40x or the oil immersion lens. Focus on one dinoflagellate, and if possible, note the two flagella characteristic of this organism (Figure 11).
2. If live dinoflagellates are available, make a wet mount and observe under the compound microscope.
3. Draw a representative dinoflagellate, either from the prepared slide or live culture.
4. Turn to Table 5 at the end of this chapter and fill in the parts of the chart that correspond to the dinoflagellates.

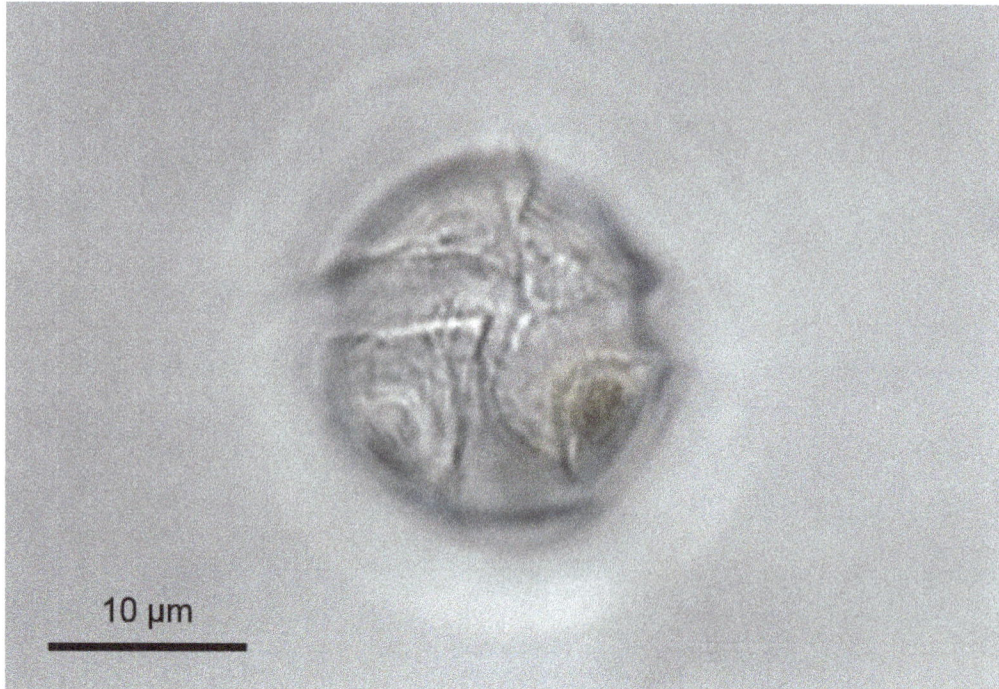

FIGURE 12. *Peridiniella danica,* **a dinoflagellate.**

F. Foraminiferans

Foraminiferans are marine organisms and are mostly heterotrophic, except for a few species that house photosynthetic symbiotic algae in their tissues. They have multichambered "shells" (tests) that can be made of different materials, depending on the species. Foraminiferans feed and move by extending fine cytoplasmic extensions through their tests (Figure 13).

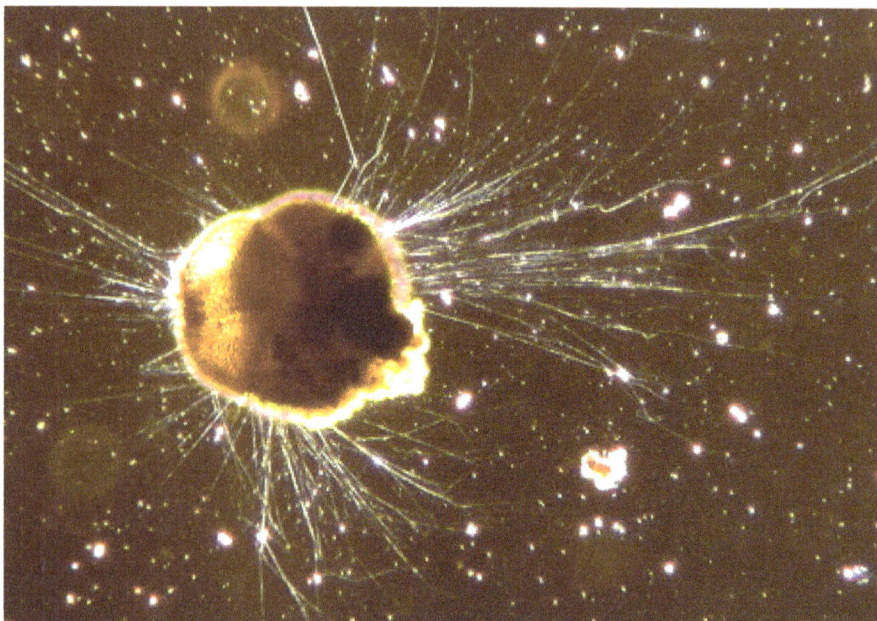

FIGURE 13. *Ammonia tepida,* **a foraminiferan.**

1. Observe a slide of foraminiferans using the compound microscope. Draw a representative fora-miniferan. Indicate the objective lens and total magnification used in viewing the specimen.
2. Turn to Table 5 at the end of this chapter and fill in the parts of the chart that correspond to the foraminiferans.

G. *Amoeba* sp. and *Chaos* sp.

Amoeba and Chaos are unicellular heterotrophs that live in freshwater environments and in moist soils. The shape of these organisms changes continuously due to the formation of pseudopodia, temporary projections of the cytoplasm and cell membrane (Figure 14). Pseudopodia are used by the organisms for movement (called "amoeboid movement") and also for feeding. During feeding, pseudopodia engulf the food item, which is then taken into the organism via phagocytosis and packaged into a food vacuole.

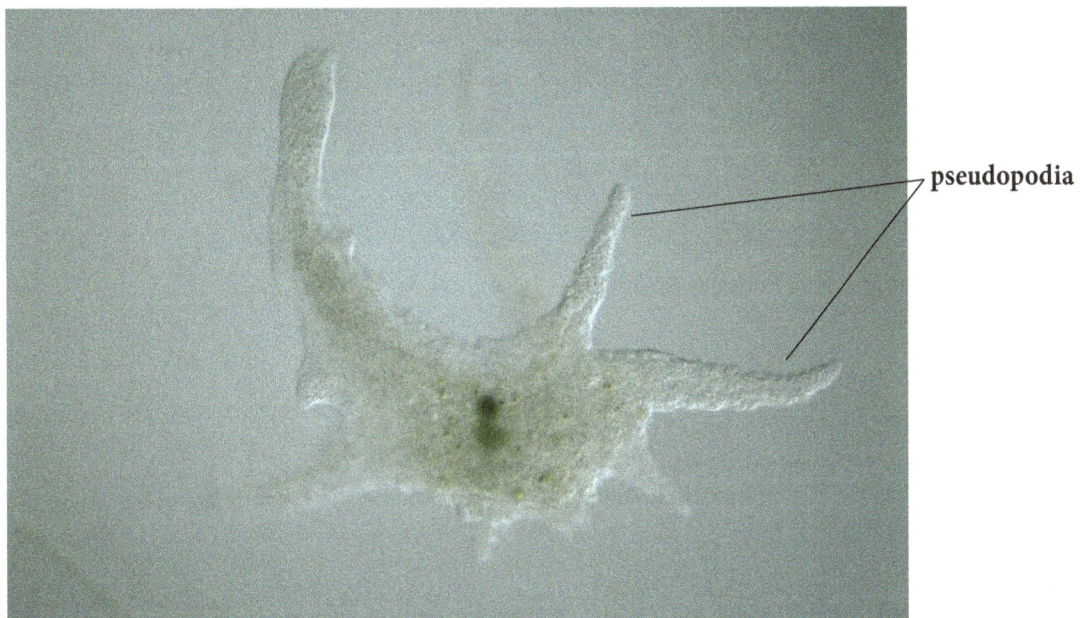

pseudopodia

FIGURE 14. *Chaos carolinensis.*

1. Observe a prepared slide of *Amoeba or Chaos* using the compound microscope. Note the pseu-dopodia, food vacuole, and nucleus (if visible).
2. Observe a live culture of *Amoeba or Chaos* using the compound microscope.
3. Draw a representative specimen (Note: The culture you are viewing may contain other live organisms in addition to *Amoeba or Chaos*.) Label the pseudopodia, food vacuole(s), and nucleus (if visible).
4. Turn to Table 5 at the end of this chapter and fill in the parts of the chart that correspond to *Amoeba*.

H. *Physarum polycephalum*

Physarum polycephalum is an example of a slime mold. Slime molds are important ecologically in that they are the main decomposers in forests. Although *Physarum* can grow to be relatively large, a single *Physarum* is actually a single cell that contains many nuclei (Figure 15).

FIGURE 15. *Physarum polycephalum*, plasmodial stage.

1. Observe the live *Physarum* specimen on demonstration under the dissecting microscope. Can you see any cytoplasmic streaming when viewed at a higher power?
2. Turn to Table 5 at the end of this chapter and fill in the parts of the chart that correspond to *Physarum*.

I. Diatoms

Diatoms are aquatic, photosynthetic organisms and they are a major component of plankton in both freshwater and marine ecosystems and are important members of aquatic food chains. Along with the dinoflagellates, diatoms are among the most abundant primary producers in the oceans. The cell walls of diatoms contain silica (SiO_2), and when diatoms die, these tests settle to the bottom of the aquatic environment and form what is known as diatomaceous earth. Diatomaceous earth is mined and is used often as an abrasive agent in polishes and toothpaste and is also used to add bulk to other products such as paint. Despite their beneficial role, several diatom genera produce a potent neurotoxin (domoic acid) that is poisonous to marine mammals and birds.

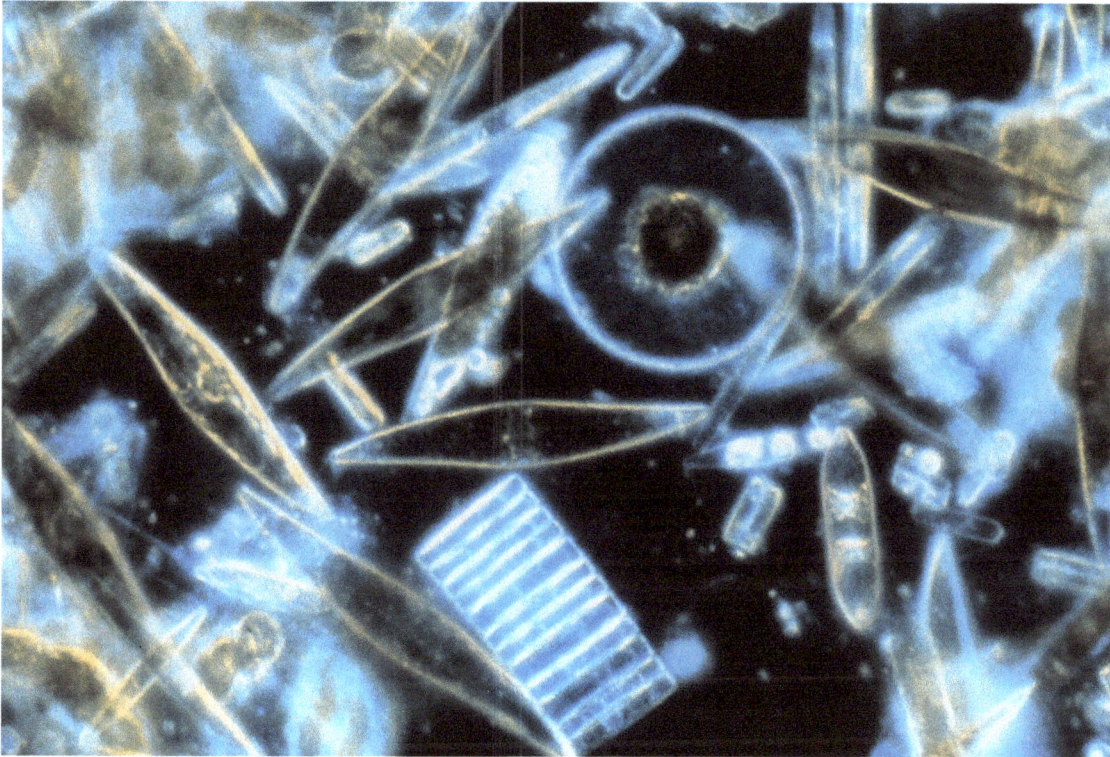

FIGURE 16. Assorted live diatoms.

1. Observe a prepared slide of diatoms using the compound microscope.
2. If live diatoms are available, make a wet mount and observe under the compound microscope. Refer to Figure 16, which shows a variety of species of diatoms, as a guide. Draw representative diatoms, from either the prepared slide or wet mount.
3. Observe the diatomaceous earth on display.
4. Turn to Table 5 at the end of this chapter and fill in the parts of the chart that correspond to the diatoms.

J. Brown Algae

Brown algae are all multicellular, photosynthetic organisms; most species are marine. They are the first organisms discussed thus far that exhibit alternation of generations. The color of brown algae comes mainly from the accessory pigment fucoxanthin.

FIGURE 17. *Laminaria* sp., a brown alga.

1. The structure of a typical brown alga includes a flat blade, a stemlike stipe, and a holdfast that attaches the alga to the substrate (Figure 17). Observe preserved and live (if available) brown algal specimens, and make any notes or sketches in the space below that will help you recognize these specimens on a laboratory practical.
2. Turn to Table 5 at the end of this chapter and fill in the parts of the chart that correspond to the brown algae.

K. Red Algae

Like the brown algae, the red algae are mostly multicellular, photosynthetic organisms, they primarily live in marine environments, and they exhibit alternation of generations. Along with green algae and land plants, they are often placed in the Plantae lineage. Their reddish color comes from phycoerythrins, red accessory pigments stored in their chloroplasts. Some red algal species form a layer of calcium carbonate on their surface.

Ascophyllum

Polysiphonia

FIGURE 18. *Polysiphonia* sp., a red alga, with the brown alga *Ascophyllum* sp.

1. Observe preserved and live (if available) red algal specimens and make any notes or sketches in the space below that will help you recognize these specimens on a laboratory practical.
2. Turn to Table 5 at the end of this chapter and fill in the parts of the chart that correspond to the red algae.

L. Green Algae

The green algae vary widely in form and in habitat: There are unicellular, colonial, and multicellular species, and green algae are found in both freshwater and marine environments. Many green algae exhibit alternation of generations.

1. Observe preserved and/or live green algal specimens available. Today you will observe the green algal genera *Chlamydomonas*, *Spirogyra*, *Volvox*, *Ulva*, and *Chara*.

a. *Chlamydomonas*

Chlamydomonas is a unicellular green alga that possesses a pair of flagella (Figure 19). It is a freshwater organism.

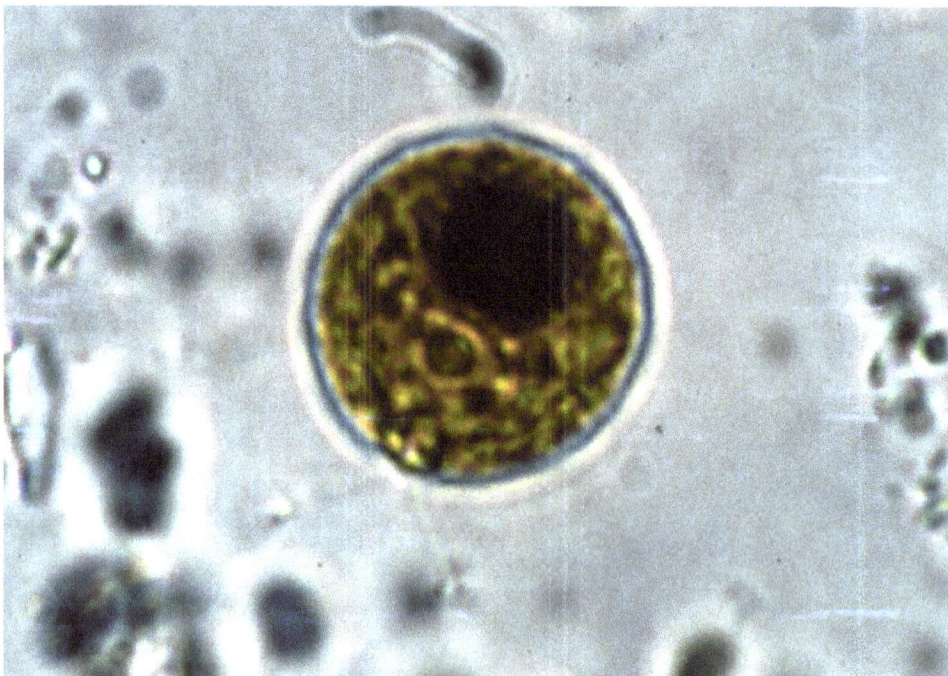

FIGURE 19. *Chlamydomonas* sp.

1. Observe a prepared slide of *Chlamydomonas* using the compound microscope. Try to locate the flagella.
2. Observe a live culture of *Chlamydomonas* using the compound microscope. Depending on the container in which the culture is living, you may be able to view the organisms in the container without having to make a wet mount; follow instructions given by your lab instructor. Draw a representative *Chlamydomonas*.

b. *Spirogyra*

Spirogyra is a filamentous, freshwater green alga that possesses a spiral-shaped chloroplast (Figure 20).

1. Observe a prepared slide of *Spirogyra* using the compound microscope. Note the chloroplast.

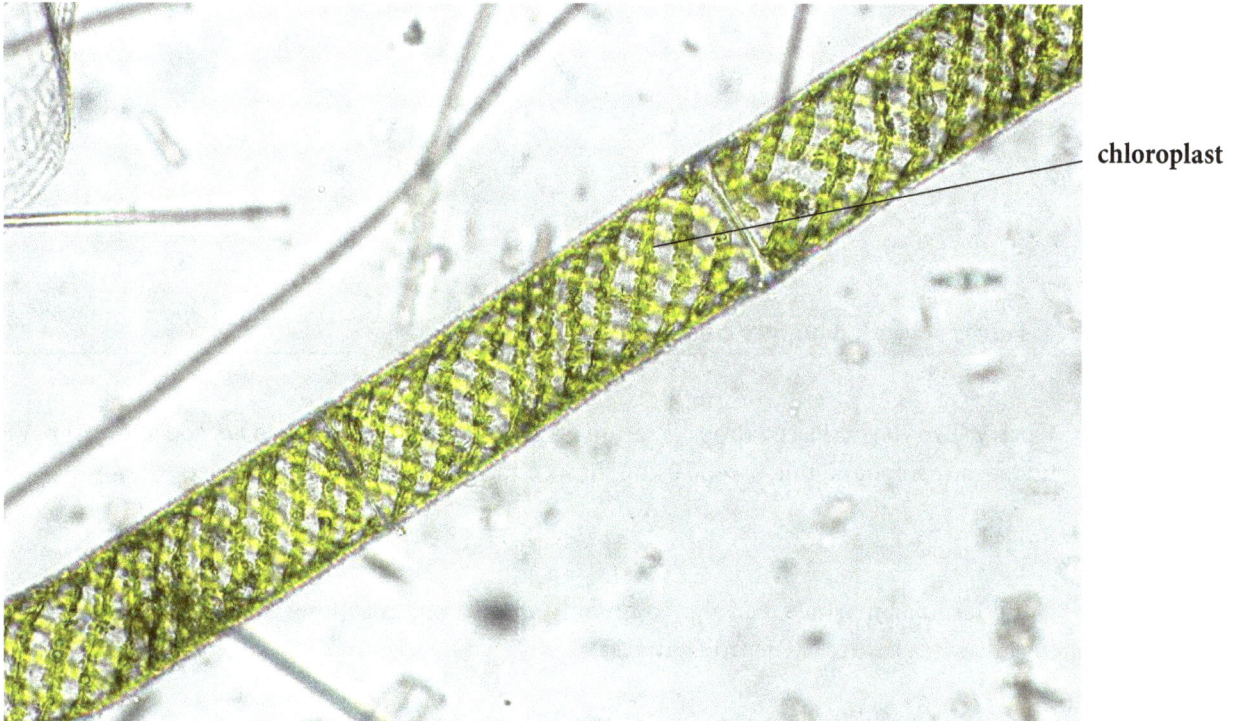

chloroplast

FIGURE 20. *Spirogyra* sp.

2. Prepare a wet mount from a live culture of *Spirogyra* and observe using the compound microscope. Draw *Spirogyra* and label the chloroplast.

c. *Volvox*

Volvox is a colonial freshwater green alga, but many sources refer to this organism as multicellular due to the cell specialization found within the organism. Colonial organisms are generally a group of individual organisms attached to one another. While there may be some specialization of cells, typically the cells of a colonial organism are able to survive on their own. The cells making up *Volvox* are unicellular, flagellated, and are very similar to *Chlamydomonas*. They are connected by thin cytoplasmic strands. There are two main cell types found in *Volvox*. One cell type is specialized for reproduction (both sexual and asexual), and another cell type, called a somatic cell, has lost the ability to divide and it functions in locomotion (Miller, 2010). This division of labor between cell types seen in *Volvox* is what leads many to refer to it as multicellular rather than colonial.

When observing *Volvox*, you may see smaller daughter colonies inside the larger organism (Figure 21). These daughter colonies were produced asexually by the reproductive cells of the parent colony. When the daughter colonies reach a certain size, the parent splits open and releases the colonies. Sexual reproduction occurs in *Volvox* via gamete production by the same reproductive cells that produce daughter colonies.

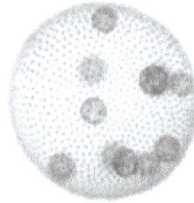

FIGURE 21. *Volvox* sp. with daughter colonies.

1. Observe live *Volvox*: Place a drop of culture onto a depression slide and do *not* add a coverslip. View under the compound microscope using the 4x and 10x objective. Draw *Volvox*.

d. Ulva

Ulva is a multicellular, marine green alga commonly found along coastlines. Because of its appearance, it is often called "sea lettuce." Refer to Figure 22.

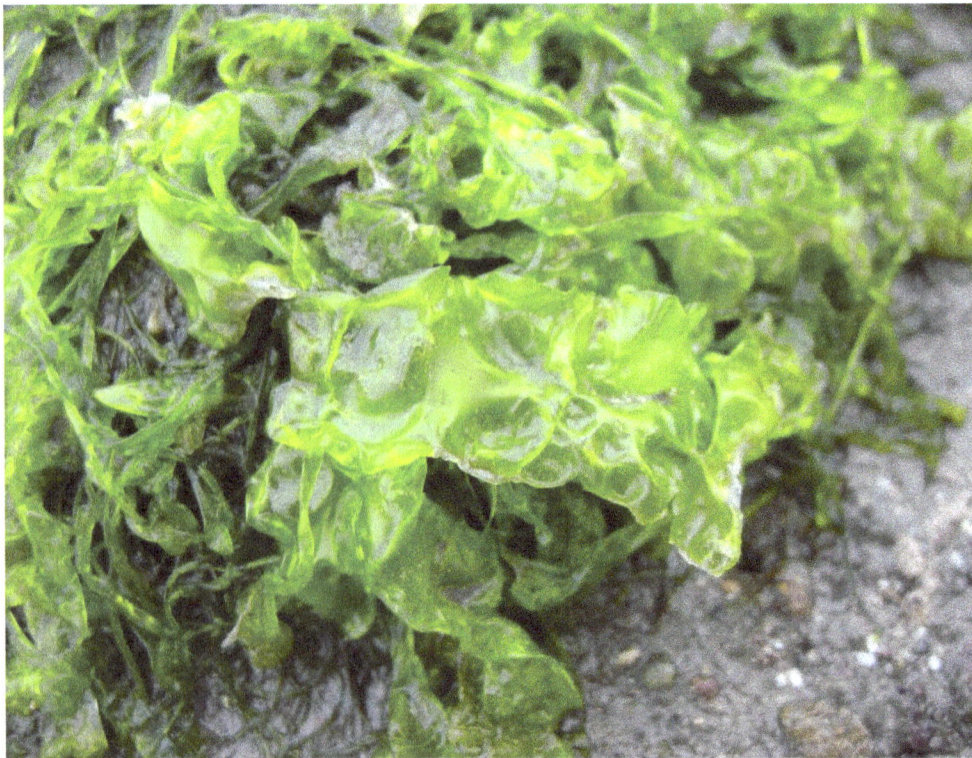

FIGURE 22. *Ulva lactuca.*

1. Observe preserved or live (if available) *Ulva* specimens.

e. Chara

Chara is a member of the charophyte algal group; charophytes are the closest living relatives of land plants.

1. Observe preserved or live *Chara* on display.
2. Turn to Table 5 at the end of this chapter and fill in the parts of the chart that correspond to the green algae.

FIGURE 23. *Chara* sp., a charophyte.

QUESTIONS

1. Compare and contrast the modes of locomotion in *Trypanosoma*, *Euglena*, *Paramecium*, and *Amoeba/Chaos*.

2. What physical feature of dinoflagellates gives them the name "dinoflagellate"? Describe any ecological importance of dinoflagellates. What is a "red tide"?

3. Describe any ecological or economic importance of diatoms.

4. List four disease-causing protists, either from today's lab or from lecture or the textbook. For each protist listed, include the name of the condition or disease it causes.

5. What are some main similarities between green algae and land plants? What groups of green algae are thought to be the closest ancestors of land plants?

Table 5. Major Lineages of Protists.

Lineage	Group	Example(s) Seen in Lab	Any ecological, economic, or evolutionary importance(s) of group or sample organism?	Visible characteristics that can help you distinguish these organisms from other protists
	Kinetoplastids			
	Euglenids			
	Ciliates			
	Apicomplexans			
	Dinoflagellates			
	Diatoms			
	Brown algae			
	Foraminifera			
	Lobose Amoebae			
	Slime molds			
	Red algae			
	Green algae			

FIGURE CREDITS

THE LAND PLANTS I: NONVASCULAR AND SEEDLESS VASCULAR PLANTS

INTRODUCTION

The three main groups of land plants are the nonvascular plants, the seedless vascular plants, and the seed plants. As with the protists, classification of the land plants into specific taxa is a topic of debate, and will therefore differ among sources. In some cases, even the taxonomic category names vary; for example, some sources use the category "phylum" and others instead use the category "division." Because of the dynamic nature of plant taxonomy, the land plants will be presented using their common names (for example, "mosses"). You may be asked to recognize a plant according to its common group name or according to its specific taxon. Your lab instructor will provide you with more details about this.

Although all of the following groups live on land, some still are tied to a watery environment. As you complete the exercises, pay attention to any characteristics that make a group more or less adapted to a terrestrial environment. Completing Table 6 at the end of the seed plants lab will be a useful exercise to help you compare and contrast the land plant groups.

ACTIVITIES

Exercise 1. The Nonvascular Plants

Nonvascular plants are plants that lack vascular tissue (xylem and phloem) to conduct water and nutrients. These plants do not possess true roots, stems, or leaves, since these structures by definition contain vascular tissue. The lack of vascular tissue to conduct substances to different parts of the plant's body limits the size of these plants; they are generally small. The lack of structural support that is provided by vascular tissue also limits the size of these plants. Additionally, they are restricted to moist habitats due to the absence of vascular tissue and also because they have swimming sperm cells.

The gametophyte generation of the nonvascular plants is the more conspicuous phase of the life cycle. Recall that the gametophyte phase in alternation of generations produces gametes via mitosis in antheridia (sperm) and archegonia (eggs). In nonvascular plants, the sperm cells are flagellated. In order for fertilization to occur, sperm must swim to the eggs, which remain in the archegonia. The resulting zygote grows into the diploid sporophyte plant, which remains attached to the gametophyte. Eventually, the adult sporophyte plant produces spores within sporangia.

Nonvascular plants include the mosses, liverworts, and hornworts. In lab, only the mosses and liverworts will be studied.

A. Mosses

1. Observe the living and preserved mosses on display. Identify the gametophyte and sporophyte generations. Note that the gametophyte is green and the sporophyte is brown (note that in younger sporophytes, the capsule is green) (Figure 24).

FIGURE 24. **Moss gametophytes and sporophytes.**

2. Obtain a prepared slide of the moss *Mnium*. This slide contains four specimens: 1) a developing filamentous gametophyte, sometimes called a protonema; 2) a longitudinal section of a gametophyte with antheridia at its tips; 3) a longitudinal section of a gametophyte with archegonia at its tips; and 4) a longitudinal section of a capsule surrounding a sporangium. Refer to Figure 25.

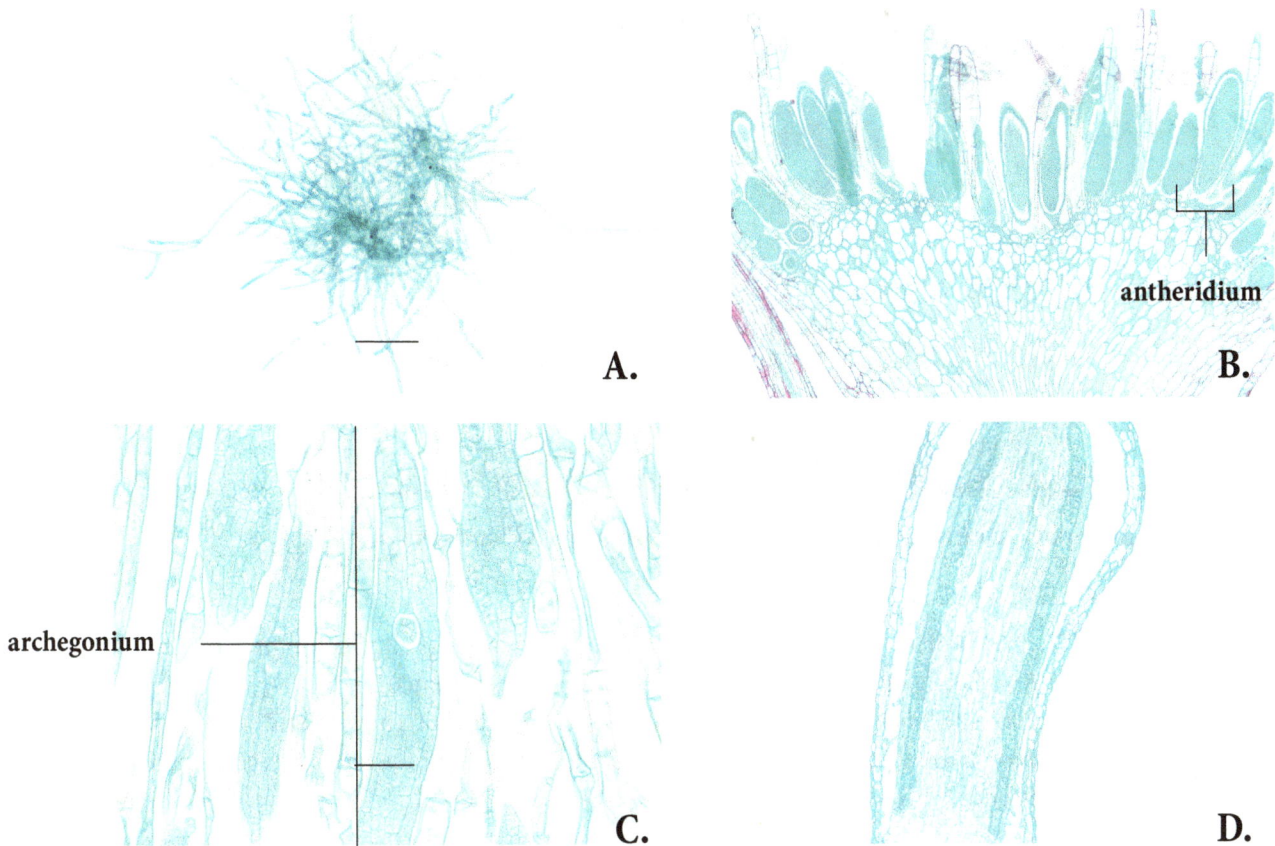

FIGURE 25. *Mnium.* A) Protonema, 25x; B) antheridia, 32x (l.s.); C) archegonium with egg, 128x (l.s.); D) capsule of sporophyte, 25x (l.s.).

A. Under the compound microscope, observe the section containing antheridia. Draw what you see, labeling an antheridium and spermatogenous tissue (the 40x objective will work best for seeing detail).

B. Observe the section containing archegonia. Draw what you see, labeling an archegonium. Also label an egg if visible, or if not visible, where an egg would be found within the archegonium (the 40x objective will work best for seeing detail).

C. Observe the longitudinal section through a capsule surrounding a sporangium, and draw what you see (the 10x objective will be best for seeing the sporangium). Label the brownish-colored spores, if present.

B. Liverworts

As with all nonvascular plants, the gametophyte is the conspicuous phase. Some liverworts produce cuplike structures called gemmae cups (Figure 26). Gemmae are cells produced asexually via mitosis in the cups; if water splashes into a gemmae cup, the gemmae splash out and get dispersed. Each gemma is capable of developing into a new gametophyte plant.

FIGURE 26. Gametophytes with gemmae cups of *Marchantia* sp., a liverwort.

1. Observe living liverworts on display, if available. Which phase(s) of alternation of generations is/are visible? Are there gemmae cups present on the living specimen(s)?

2. Observe preserved specimens of the liverwort *Marchantia* on display. The antheridia and archegonia of *Marchantia* are located on structures called antheridiophores and archegoniophores (Figure 27). The ends of these structures, receptacles, are where archegonia and antheridia may be found.

3. OPTIONAL EXERCISE: Observe a prepared slide of a longitudinal section through an antheridiophore and male receptacle of *Marchantia*. Draw what you see and label as directed by your instructor.

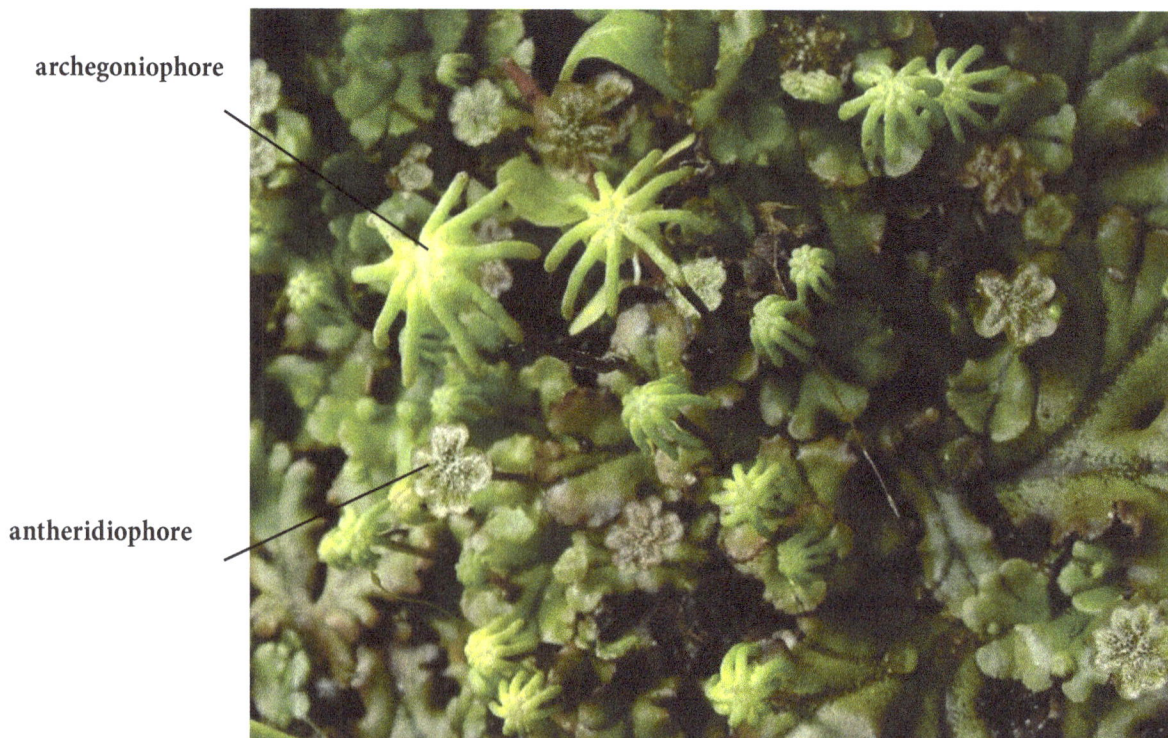

FIGURE 27. Antheridiophores and archegoniophores of *Marchantia* sp., a liverwort.

4. OPTIONAL EXERCISE: Observe a prepared slide of a longitudinal section through an archegoniophore and female receptacle of *Marchantia*. Draw what you see and label as directed by your instructor.

Exercise 2. The Seedless Vascular Plants

Seedless vascular plants, as the name suggests, do not produce seeds but do possess vascular tissue, xylem and phloem. Phloem transports sugars, while xylem transports water and minerals such as nitrogen. Seedless vascular plants do have leaves, roots, and stems, although in some seedless vascular plants, these are modified (for example, ferns have modified leaves called fronds). Unlike nonvascular plants, the gametophyte of seedless vascular plants is usually independent of the sporophyte, and the sporophyte is the more conspicuous phase.

The presence of vascular tissue in this group enables these plants to grow taller than the nonvascular plants, since the vascular tissue provides support, in addition to carrying water and nutrients to those parts of the plant that are not in direct contact with water. However, similar to the nonvascular plants, these plants possess flagellated (swimming) sperm cells. Thus, they are still tied to a moist environment.

This group includes club mosses (which are not mosses), whisk ferns (which are not ferns), horsetails, and ferns. In the following lab activities, you will observe live and/or preserved examples of seedless vascular plants, focusing on the club mosses, horsetails, and ferns.

A. Club Mosses

Club mosses, despite their name, are not mosses. They have true roots, small leaves called microphylls, and stems. *Lycopodium* and *Selaginella* are examples of club mosses.

FIGURE 28. *Lycopodium* **sp.**

FIGURE 29. Strobili of *Selaginella* sp.

1. Observe preserved specimens of *Lycopodium* on display. These are sporophyte plants. Note the club-shaped structures present at the tips of some of the specimens. These structures are clusters of sporangia called strobili (sing. = strobilus).

2. Observe live (if available) and preserved *Selaginella*. The plants you are observing are sporophytes. *Selaginella* also possesses strobili (Figure 29); locate these structures on a live or preserved *Selaginella* specimen.

3. After examining *Selaginella* and locating strobili, observe a prepared slide of a longitudinal section of a *Selaginella* strobilus to observe the sporangia (Figure 30). You will see megasporangia, which produce large megaspores, and microsporangia, which produce smaller microspores. Megaspores are spores that will develop into female, egg-producing gametophytes. Microspores develop into male, sperm-producing gametophytes. Draw what you see, labeling the megaspores and microspores.

4. In the space below, make any notes or drawings that will help you identify club mosses on a lab practical. What are some *visible* characteristics that can help you tell the difference between *Selaginella* and *Lycopodium* and a true moss looked at in the previous section?

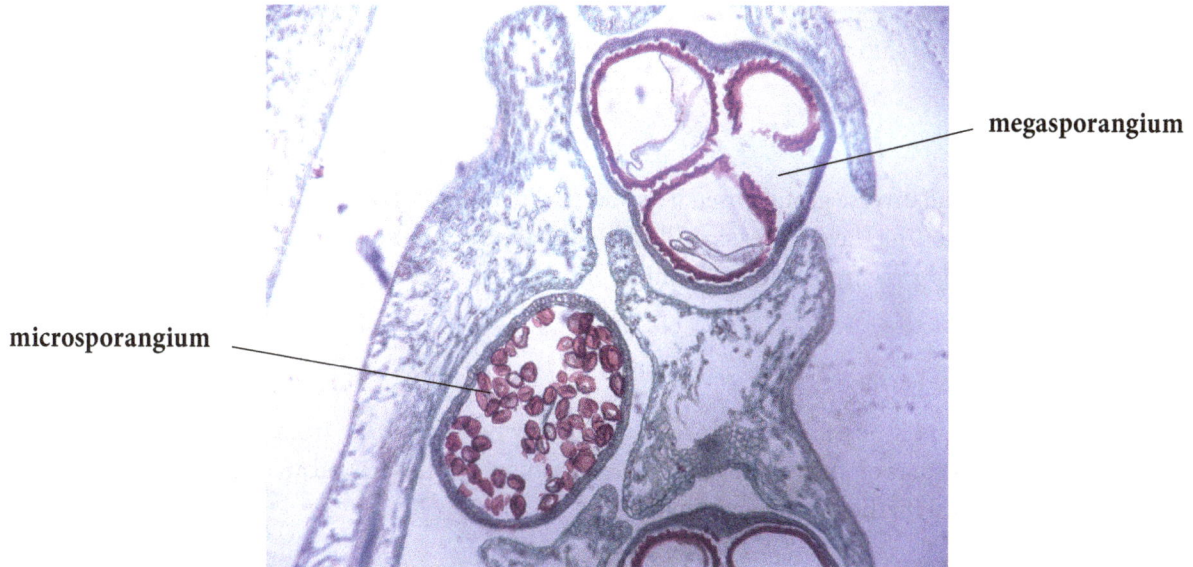

FIGURE 30. Longitudinal section of a *Selaginella* strobilus, showing a megasporangium and microsporangium.

B. Horsetails

Horsetails are found in wet habitats and get their name from the brush-like leaves that are attached to the stems at regular intervals (Figure 31). Some stems have visible strobili at their tops. *Equisetum* is the only genus of horsetails that is living today.

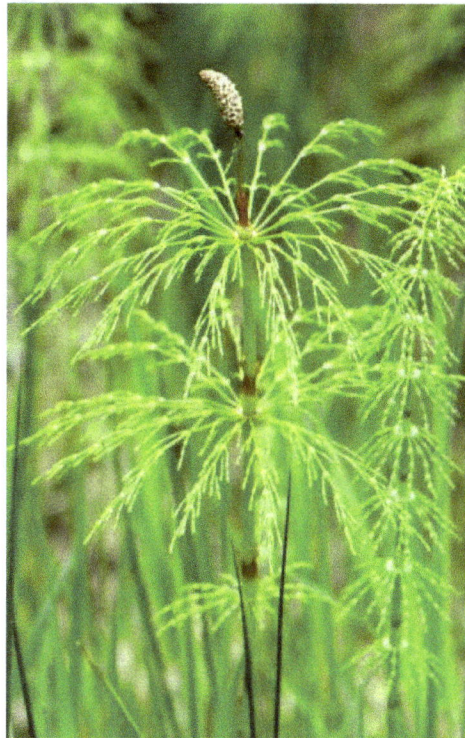

FIGURE 31. *Equisetum* sp.

1. Observe live and preserved *Equisetum* specimens. Locate the strobili. Recall that strobili were also visible in the club mosses. What is a strobilus? What are produced by a strobilus?

C. Ferns

The most abundant and diverse group of seedless vascular plants is the ferns. They vary in size and live in moist environments. The leaves of ferns, called fronds, are the largest of the seedless vascular plants (Figure 32). The fronds grow from underground stems called rhizomes.

1. Observation of Live Ferns

Identify the fronds, and if possible, any rhizomes that may be protruding from the soil. Some of the fronds may have dark spots on their undersides. These spots are called sori and are clusters of sporangia. These will be discussed further in the next exercise.

— sorus

FIGURE 32. Closeup of sori (sing., sorus). Each sorus is a cluster of sporangia.

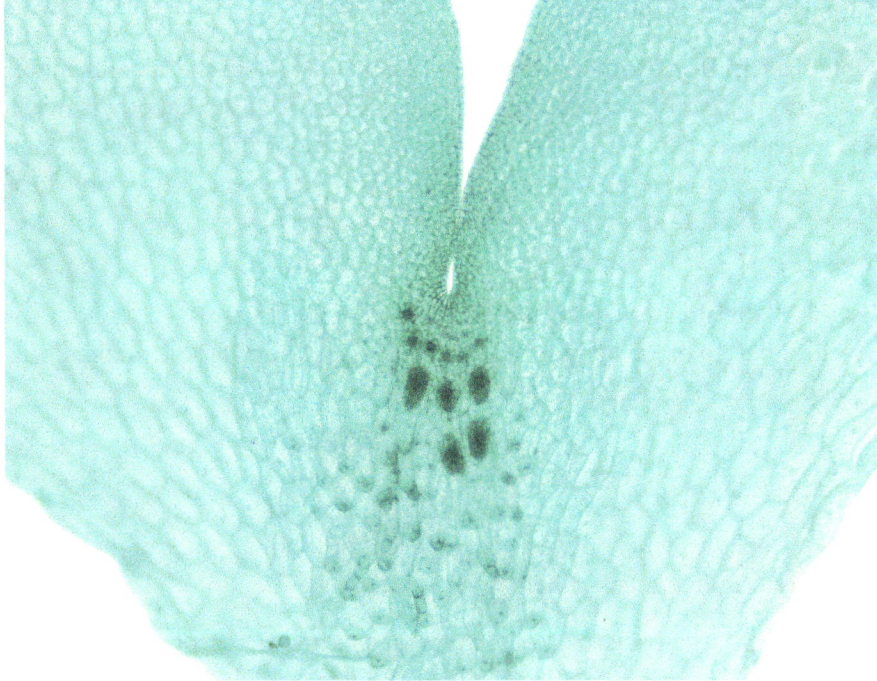

FIGURE 33. **Fern gametophyte with archegonia, 45x.**

2. The Fern Life Cycle

As with all seedless vascular plants, the fern sporophyte is the most conspicuous phase of the life cycle. The sporangia, when present, are found in clusters on the undersides of the fronds. Each cluster of sporangia is called a sorus (pl., sori). Spores, upon their release from a sporangium, are capable of developing into gametophytes when the conditions are favorable. The fern gametophyte is a small, green, heart-shaped plant. Fertilization occurs via sperm (produced in antheridia) swimming to eggs within archegonia on the gametophytes. The resulting zygote remains in the archegonium and develops into a young sporophyte.

In order to better understand the fern life cycle, you will be observing the different stages of and structures involved in fern reproduction. As you complete the activities, it will be helpful to have a diagram of the fern life cycle available.

A. Observation of fern sporophytes

1. Obtain a fern frond with sori from the designated area in the lab and cut off a small part of the frond for observation; return the rest of the frond so that others may use it.
2. Examine the sori on your frond piece under a dissecting microscope at your bench.
3. Place a drop of glycerol on one of the sori. You may eventually be able to see individual sporangia rupture and release spores.

What process of cell division resulted in the production of these spores? _____
What is the ploidy level (1n or 2n) of these spores? _____ Of the sporangia? _____
Of the fern frond? _____

B. Observation of fern gametophytes

(NOTE: Fern gametophyte plant bodies are often called "prothallia"; some of the slides viewed in the exercises may use this term instead of gametophyte.)

1. Using tweezers, place a live fern gametophyte on a clean microscope slide. Some of the gametophytes might be clumped together, but try to isolate a single plant for viewing. Do not add a coverslip.
2. At your bench, view the gametophyte under a dissecting microscope. Note its shape and color. Hairlike structures called rhizoids anchor gametophytes in the soil; are rhizoids visible? See if you can identify the general location of the archegonia or antheridia. Often, a fern gametophyte will contain either, but not both, of these reproductive structures. Fern spores have the potential to develop into either male or female gametophytes (Banks, 1999).
3. Add a drop of water to the gametophyte and make a wet mount. Gently press on the coverslip to flatten the gametophyte. View this slide under the compound microscope with the 4x objective in place. Once you are able to determine the general location of antheridia and/or archegonia, increase the magnification and focus on the antheridia. Sometimes it is possible to see the motion of sperm swimming from the antheridia. Are you able to see this?
4. Obtain a prepared slide of a fern gametophyte containing antheridia and archegonia. These slides often contain two gametophytes: one with visible archegonia and one with visible antheridia. Draw a fern archegonium and antheridium and label the egg and the sperm/spermatogenous tissue.

What process of cell division results in the production of these egg and sperm by gametangia? _____

What is the ploidy level (1n or 2n) of eggs and sperm? _____ Of the cells of an archegonium? _____ An antheridium? _____

   C.   Observation of a new sporophyte

As seen in the mosses, fertilization in the ferns occurs as a result of sperm swimming to the eggs, which remain in the archegonia. Young sporophyte ferns therefore grow out of the archegonia on gametophytes (Figure 34).

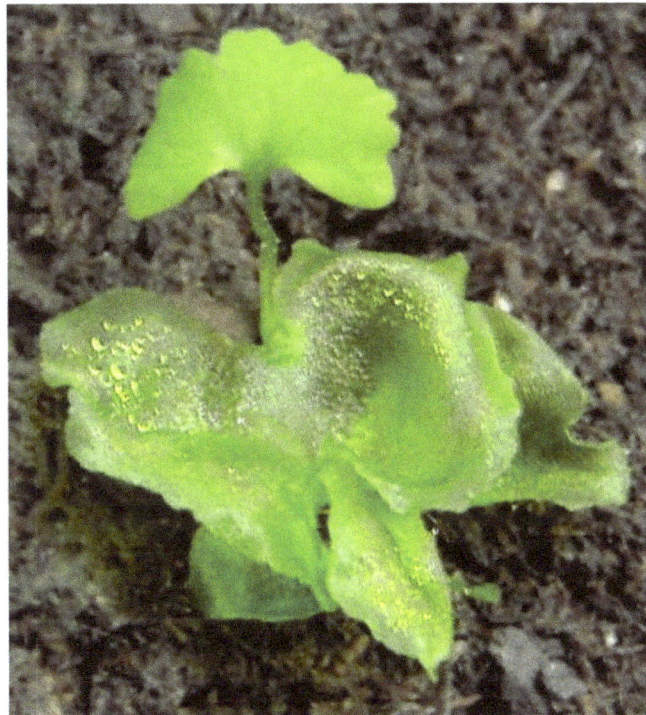

FIGURE 34. Embryo fern sporophyte growing from gametophyte.

   1.   If available, observe living young sporophytes growing from gametophytes that are on display.

QUESTIONS

1. State the ploidy level (1n or 2n) of the following:
   antheridium_____
   archegonium_____
   sporangium_____
   egg_____
   sperm_____
   spore_____

2. Fill in the blanks using the following terms: sporangia, mitosis, meiosis, antheridium, archegonium, gametangia, gametophyte, sporophyte.

   In alternation of generations, gametes are produced by structures called_____
   via the cellular division process_____. Eggs are produced by
   _____and sperm are produced by_____. Spores are
   produced in structures called_____via the cellular
   division process_____

3. Compare and contrast the life cycles of the mosses and ferns. How are they similar? How are they different? In the space below, sketch the life cycle of the moss and the life cycle of the fern, indicating which phase of alternation of generations is the conspicuous phase in the life cycle.

4. What makes the seedless vascular plants more adapted for living on land than nonvascular plants?

5. What features tie nonvascular and seedless vascular plants to a watery environment?

6. Turn to Table 6 at the end of the seed plant exercises and fill in the portions that pertain to the land plants studied thus far.

FIGURE CREDITS

THE LAND PLANTS II: SEED PLANTS

INTRODUCTION

There are two main groups of seed plants: gymnosperms and angiosperms. Gymnosperms include the pines, cycads, ginkgoes, junipers, and redwoods, among others, and the angiosperms are the flowering plants. The seed plants are completely adapted to a terrestrial environment. The embryos of seed plants are housed inside seeds, which provide the initial nutrition for the embryos and protect them from desiccation. Seeds also provide a means of dispersal of the embryos from the parent plant.

The evolution of pollen grains in this group also makes seed plants better suited for life on land than the nonvascular plants or seedless vascular plants. Pollen grains are male gametophytes that produce sperm nuclei that fertilize eggs. Recall that both nonvascular plants and seedless vascular plants require a moist environment (for example, a sheet of water or a raindrop) for their flagellated sperm to swim to eggs within archegonia. In the seed plants, water is not required for the sperm to reach eggs: pollen grains are carried by wind and/or animals to female gametophytes, at which point the sperm nuclei are delivered to the eggs and fertilization occurs.

The details of the seed plant life cycles will be presented in lecture, but a general overview will be presented here to help clarify reproduction in this group and as a reference for completing today's lab activities. Although the seed plant life cycles are fairly complicated, it is useful to remember that these plants, like others studied thus far, carry out alternation of generations. The sporophyte generation is the conspicuous phase and the gametophytes of seed plants are much smaller than those of the nonvascular plants and seedless plants. The female gametophytes of seed plants are microscopic, and as mentioned, the male gametophytes are pollen grains.

The following diagram illustrates the main stages and structures found in the gymnosperm and angiosperm life cycles.

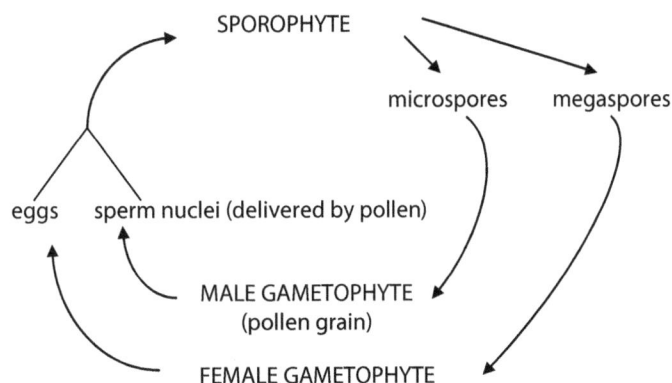

Egg development in seed plants

Gymnosperms and angiosperms possess structures called ovules. In gymnosperms such as the pines, ovules are located on female pine cones; they are not enclosed (gymnosperm means "naked seed"). In angiosperms, ovules are enclosed within one or more ovaries (angiosperm means "seed enclosed in a vessel"). Ovaries are the base of the carpel, the entire female floral part. Each ovule consists of an integument (coating) surrounding a megasporangium. The megasporangium produces a haploid megaspore via meiosis, which develops into the female gametophyte, a microscopic structure completely enclosed by the ovule. The egg is produced by the female gametophyte via mitosis; the egg and gametophyte remain inside the integument. To summarize the events occurring within the integument:

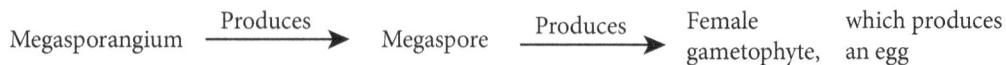

$$\text{Megasporangium} \xrightarrow{\text{Produces}} \text{Megaspore} \xrightarrow{\text{Produces}} \text{Female gametophyte,} \quad \text{which produces an egg}$$

Pollen grain and sperm development in seed plants

Pollen grains develop in microsporangia. In gymnosperms such as the pines, microsporangia are located in male pine cones. In angiosperms, microsporangia are located in the anthers of flowers. Anthers are located at the tips of the male floral parts called stamens. Within the microsporangia, haploid microspores are produced via meiosis. Each microspore develops into a male gametophyte (pollen grain). The pollen grains are then released, and when they travel via wind or animal to a female reproductive structure, the pollen grains produce sperm nuclei via mitosis, which are then delivered to the egg in the ovule. To summarize the events occurring within the microsporangium:

$$\text{Microsporangium} \xrightarrow{\text{Produces}} \text{Microspore} \xrightarrow{\text{Produces}} \text{Male gametophyte (pollen grain),} \quad \text{which produces sperm nuclei once in contact with female reproductive structure}$$

Fertilization in seed plants

In gymnosperms, wind delivers the pollen to the ovules. In pines, evaporation of sticky sap on the female cones helps draw pollen grains to the eggs in the ovules. In angiosperms, pollen grains are delivered from the anthers to the sticky tips of the carpels (stigma) via wind or animals. The pollen grains then develop pollen tubes that deliver the sperm nuclei to the eggs in the ovules. The transfer of pollen to the female gametophyte is called pollination.

After fertilization, an ovule develops into a seed. The former integument develops into a seed coat, the fertilized egg (zygote) develops into an embryo sporophyte plant, and the tissue surrounding the embryo functions as the initial source of nutrition for the embryo once the seed germinates.

Upon fertilization in some angiosperms, the ovary expands and matures into a fruit.

ACTIVITIES

Exercise 1. Gymnosperms

The representative gymnosperms we will look at in lab are the ginkgoes and the pines.

A. Observe any live or preserved ginkgoes and pines on display.

B. Study the different stages of the pine life cycle by completing the following exercises. As you observe the stages, try to match each with a pine life cycle diagram from your lecture notes or text.

1. Observe the male pine cones on display. The scales on male cones are the locations of the microsporangia (Figure 35).

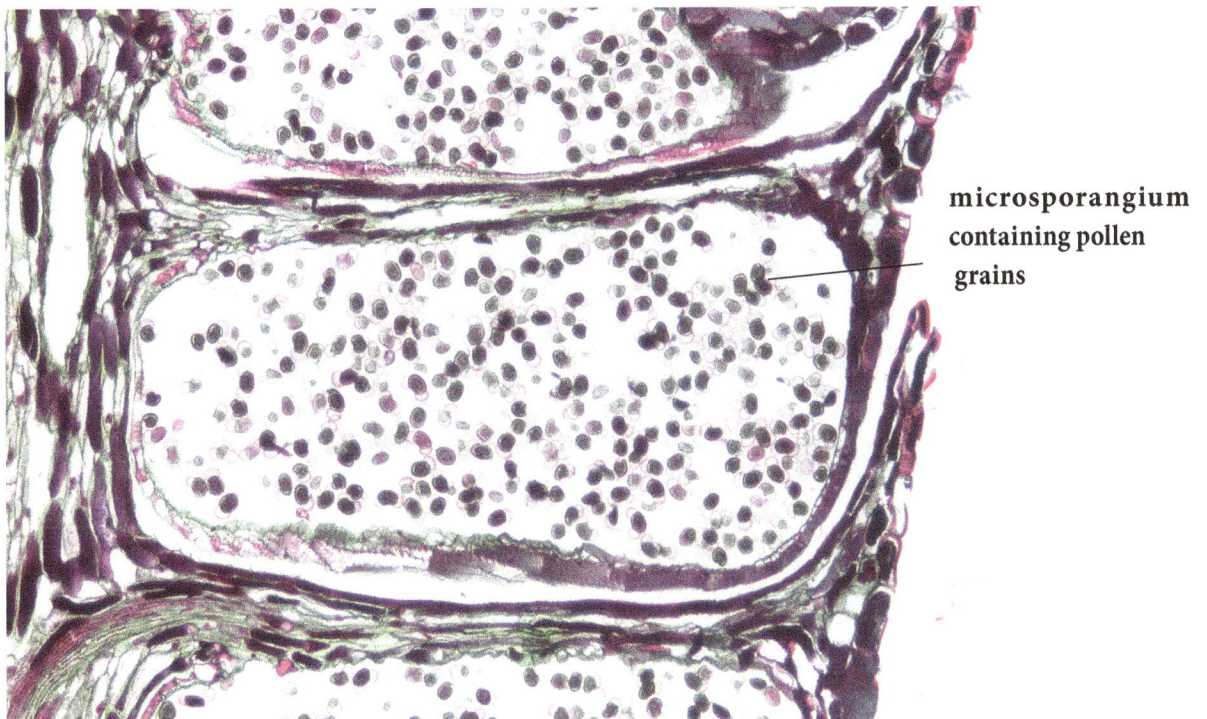

microsporangium containing pollen grains

FIGURE 35. Microsporangia of male strobilus, *Pinus* sp. (l.s.), 45x.

2. Obtain a prepared slide of a male pine cone and observe under a compound microscope at your bench. Focus on one cone scale and locate the microsporangium containing either microspores or pollen grains, depending on the developmental stage of the cone. Draw what you see and label the microsporangium and pollen grains or microspores.

3. Obtain a prepared slide of pine pollen and observe under a compound microscope at your bench. Draw what you see.

4. Observe the preserved pine ovules on display. Then, observe any female pine cones on display. Try to locate the areas on the scales where ovules once were.

5. Observe the prepared slide on demonstration showing a longitudinal section through a female cone (Figure 36). You should see the cone scales and the location of the ovules. It can be difficult to identify specific structures within an ovule, but try to locate any of the following: megasporangium,

megaspore, egg. Different ovules may be in different stages of maturation, so you may have to look at different ovules to locate them. Draw what you see, labeling any structures you draw.

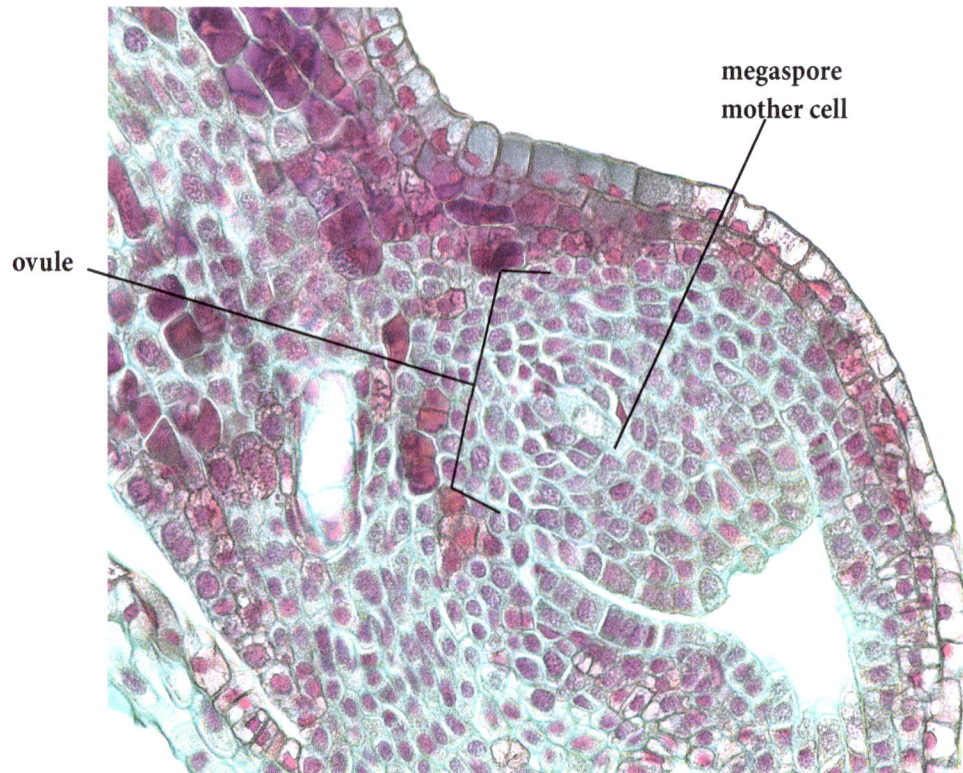

FIGURE 36. Ovule on female cone scale of *Pinus* **sp. (l.s.), showing megaspore mother cell, 43x.**

6. Observe the pine "nuts" on display. Pine "nuts" are actually pine seeds; the pine "nuts" purchased in a grocery store are missing the integument. Look at the cut pine seed on demonstration under a dissecting microscope. When sliced longitudinally, the embryo sporophyte is visible inside.

Exercise 2. Angiosperms

A. Observe the flower model on demonstration and Figure 37 and identify the following structures:

1. **Sepals:** the outermost floral parts; these are either green in color or are similar in appearance to petals; the sepals completely enclose a flower when it first appears as a bud
2. **Petals:** floral parts just inside of the sepals
3. **Stamens:** male reproductive structures of the flower
4. **Anthers:** the tips of the stamens; sites of pollen grain production
5. **Carpel:** female reproductive structure of the flower; composed of ovary, style, and stigma
6. **Ovary:** base of the carpel, ovules are located inside
7. **Style:** elongated portion of the carpel extending from the ovary
8. **Stigma:** tip of the style where pollination occurs
9. **Receptacle:** base at which all floral parts are attached

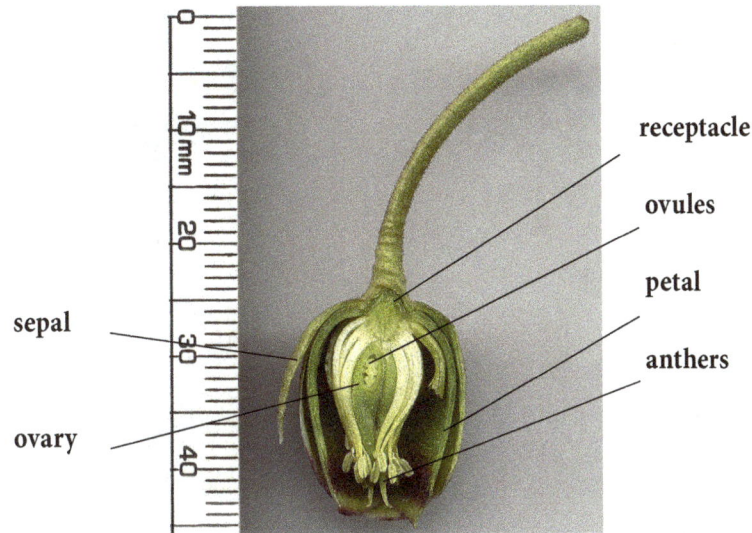

FIGURE 37. Longitudinal section of *Helleborus foetidus,* a member of the buttercup family.

B. Dissect a live flower as follows. If more than one flower is available, dissect all types.

1. Obtain a dissecting tray, flower, and scalpel or razor blade.
2. Locate the sepals, the outermost floral parts. Recall that these may look identical to the petals.
3. Carefully remove the sepals and locate the petals.
4. Locate the anthers and carpel(s). Using the scalpel or razor blade, make a longitudinal slice through the carpel. Observe the ovary under a dissecting microscope. Locate the ovules. If you are having difficulty locating these, try increasing the magnification and/or changing the light intensity.

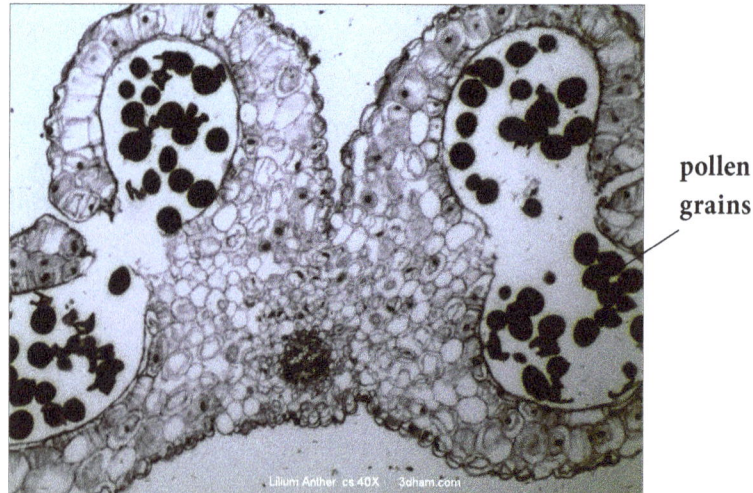

FIGURE 38. Cross section of *Lilium* sp. anther.

C. Study the different stages of the angiosperm life cycle by completing the following exercises. As you observe the stages, try to match each with an angiosperm life cycle diagram from your lecture notes or text.

1. Observe the prepared slide on demonstration showing a cross section through the anthers of a lily (*Lilium* sp.). Identify the microspores or pollen grains (depending on the stage of development) within the anthers' chambers (Figure 38). Note that, depending on the slide preparation, a slice may have also been made through the carpel of the lily. If this is the case, it would be located in the center of the slide preparation, encircled by the anthers.

2. Observe the prepared slide on demonstration showing a cross section through the ovary of *Lilium* (Figure 39). Depending on the stage of development, different structures will or will not be visible in the slide. Read the description accompanying the display to determine the stage of development being shown. In what stage of development are the ovules? Have the eggs been fertilized?

3. Pollen grains of certain flowers can be coaxed to grow pollination tubes in the presence of certain chemicals. Using the pollen from the flower(s) designated by your lab instructor, follow the instructions below for observing pollen tube growth.

   A. Obtain a Petri dish, piece of filter paper, dropper bottle of water, a dissecting probe, a glass slide, and a dropper bottle containing pollen tube growth medium.

   B. Open the Petri dish, place the filter paper in the dish, and cut to fit if necessary. Add a few drops of water, just until the paper is moist. Replace the lid.

   C. Place 1 or 2 drops of pollen tube growth medium on the glass slide. To the medium, add a few paintbrush bristles from the designated area in the lab. This will prevent the pollen grains from being crushed.

   D. Touch the dissecting probe to the anther of the flower type indicated by your lab instructor. Then touch the probe to the medium such that some of the pollen grains wash into the medium. Gently add a coverslip and place the slide into the Petri dish on top of the filter paper. Place the dish in the designated area.

   E. After 45–60 minutes, observe the slide under the compound microscope. Draw what you see and label a pollen grain and pollen tube.

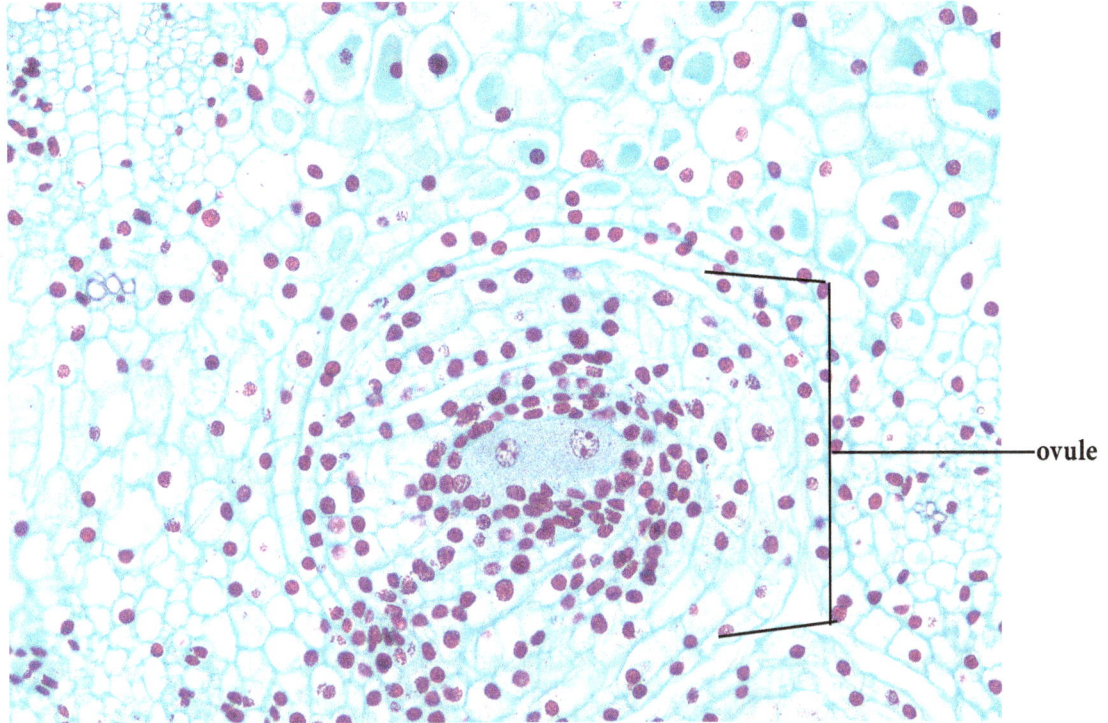

FIGURE 39. Cross section through an embryo sac within an ovule of *Lilium* sp., 106x.

Pollen grains of flowering plants contain two nuclei, one of which forms the pollen tube, and the other of which divides by mitosis to form two sperm nuclei. These sperm nuclei carry out "double fertilization." Consult your lecture notes and text for more information on double fertilization.

As mentioned, fruits form in many angiosperms after fertilization occurs. As the ovules of a flower mature into seeds, the ovary matures into a fruit. There are many types of fruits. This variety among fruits is caused by many factors, including differences in the number of ovaries present on the flower and the position of the ovaries on the flower. In addition, some fruits are the result of the fusion of multiple parts of one flower or of many flowers. Fruits aid in the protection of the seed and also provide a means for seed dispersal. Some fruits are carried by the wind away from the parent plant, while others are eaten by animals and deposited in feces away from the parent plant.

4. Observe any fruits on display and contrast these with any vegetables on display. What is the difference between a fruit and a vegetable? (Hint: What do fruits possess that vegetables do not?)

QUESTIONS

1. How are fruits important in seed dispersal?

2. Technically, what is a pollen grain? In terms of reproduction, what advantage do pollen grains give the seed plants over the seedless vascular plants or nonvascular plants?

3. Explain how the evolution of the seed contributed to making the seed plants completely adapted for life on land.

4. Complete Table 6 that follows.

TABLE 6. A Comparison of Main Groups of Land Plants.

	Mosses	Liver-worts	Club Mosses	Ferns	Horsetails	Gymnosperms	Angiosperms
gametophyte or sporophyte conspicuous/ dominant?							
is vascular tissue present?							
is water required for fertilization?							
are seeds present?							
is pollen produced?							
are fruits produced?							

FIGURE CREDITS

FUNGI

INTRODUCTION

The fungi are the most important decomposers on the planet. They are saprophytic, meaning they feed on dead organic matter. Unlike organisms that ingest their food, saprophytes secrete digestive enzymes onto the substrate and then absorb the products of the breakdown.

There are unicellular and multicellular fungi. The body of a multicellular fungus is made of filamentous structures called hyphae. Hyphae are either septate, with perforated walls dividing the nuclei, or coenocytic, with no septa. The hyphae of a fungus are often organized into a mass called a mycelium. Another characteristic of most fungi is the presence of chitin, a polysaccharide, in their cell walls.

Fungal life cycles are complex and vary between groups. Some fungi exhibit both sexual and asexual phases, some possess only a sexual phase, and some species have no known sexual phase. One characteristic common to all fungal life cycles is the formation of sexual and/or asexual haploid spores, which function in dispersal of the fungus. Spores remain dormant until environmental conditions are appropriate for germination.

In addition to being important ecologically as decomposers, fungi are also important economically. For example, yeasts are used in making bread and beer, fungi such as mushrooms are eaten, some molds are used in cheese-making, and some fungi produce antibiotics. In addition, some species cause disease.

The three fungal phyla that will be focused on in lab are Zygomycota, Ascomycota, and Basidiomycota.

ACTIVITIES

Exercise 1. Observation of Representative Fungi

A. Phylum Zygomycota

The zygomycetes get their name from their formation of structures called zygosporangia during reproduction (Figure 40). Zygosporangia form when hyphae fuse or unite (*zygo-* = yoke; pair) (Freeman, 2011). Examples of zygomycetes you will observe in lab are *Rhizopus stolonifer* (black bread mold) and *Pilobolus* (dung fungus).

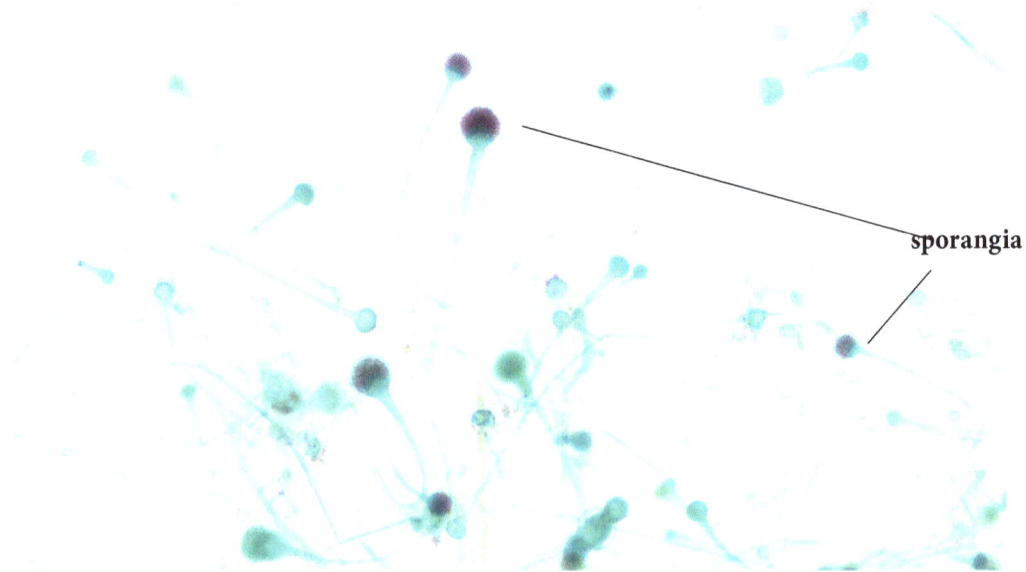

FIGURE 40. *Rhizopus stolonifer*, showing sporangia, 91x.

1. *Rhizopus stolonifer*

Although commonly called black bread mold, *Rhizopus stolonifer* also grows on other substrates, including some fruits.

    A. Observe live *Rhizopus stolonifer* cultures on demonstration and identify the following structures if present: hyphae, zygosporangia, sporangia.

    B. Obtain a prepared slide of *Rhizopus stolonifer* and observe under a compound microscope at your bench. Depending upon the individual slide, different structures might be visible, including zygosporangia, sporangia, hyphae, and spores. Draw what you see and label any visible structures.

2. *Pilobolus* sp.

Pilobolus sp. is a zygomycete that grows on animal dung. The sporangia of *Pilobolus*, located at the tips of hyphae growing upward out of the substrate, are discharged from the hyphae when exposed to bright sunlight. This causes dispersal of the sporangia and therefore of the spores inside them.

    A. Observe the *Pilobolus* culture on demonstration. Be sure to keep the light source on the culture at a low intensity. Turn the light off when you are finished.

B. Phylum Ascomycota

The ascomycetes include the cup fungi, truffles, molds such as *Penicillium*, and unicellular yeasts, among others. They get their name from the fact that many multicellular ascomycetes form saclike structures called asci (singular = ascus; *asco-* = sac, bag) (Figure 41). Within asci, haploid ascospores are produced via meiosis. For more details on reproduction in ascomycetes, consult your lecture notes and textbook.

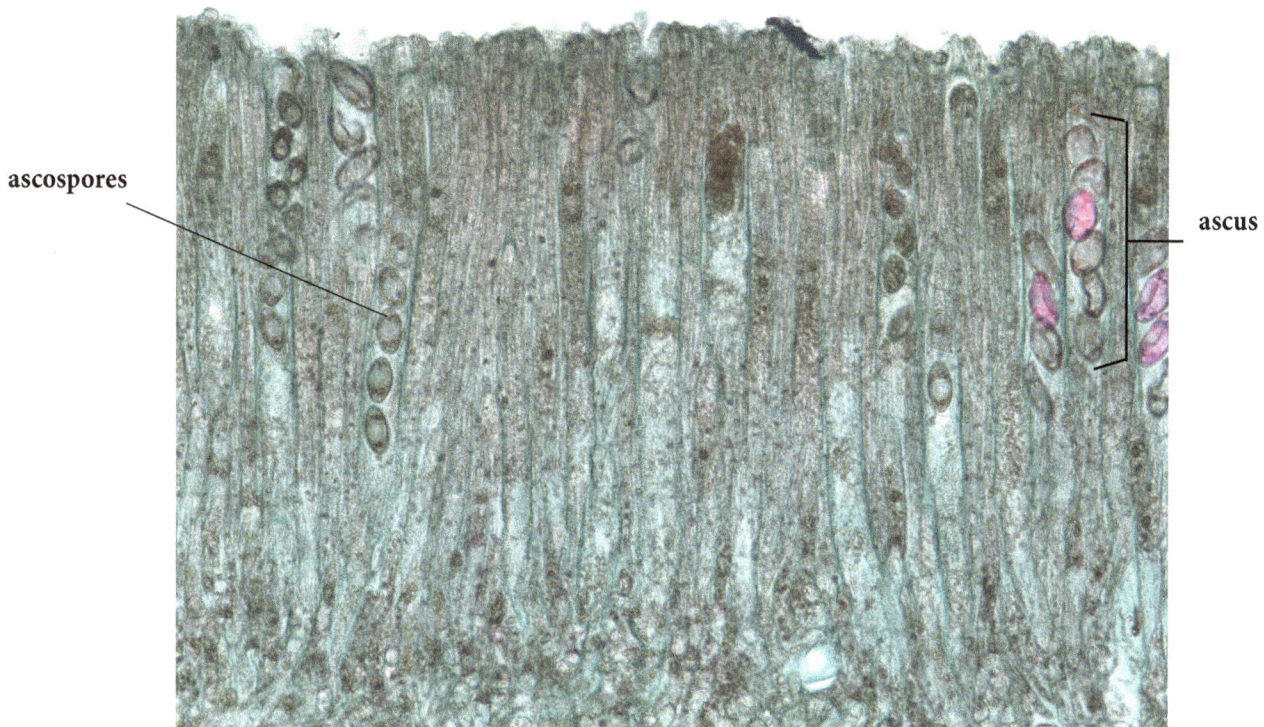

ascospores

ascus

FIGURE 41. Asci of *Peziza* sp., showing ascospores, 450x.

1. *Peziza* sp.

Peziza is a cup-shaped ascomycete that grows on wood.

  A. Observe preserved *Peziza* specimens, if available, and then obtain a prepared slide of a longitudinal section of a *Peziza* ascocarp. An ascocarp is a structure in which asci are located (it is the cup-shaped fungal body). Observe the slide under a compound microscope and draw what you see. Label an ascus and ascospores.

2. *Penicillium*

Penicillium is a genus of molds that is very important economically, as certain members make penicillin and some are used in making cheeses such as Roquefort and camembert (*P. roqueforti*, *P. camemberti*). The bluish color of this mold is due to conidia, haploid asexual spores. The conidia are found in chains attached to the structure that produces them, the conidiophore (Figure 42).

  A. Observe the *Penicillium* culture on display under the dissecting microscope. Focus closely on the culture and try to locate the conidia.

  B. After observing the live culture, observe a prepared slide of *Penicillium* under the compound microscope. Try to locate conidiophores and conidia.

  C. Observe the display showing Roquefort cheese underneath the dissecting microscope. What can be seen in the bluish areas of the cheese?

FIGURE 42. *Penicillium* sp.

3. Yeasts

Yeasts are unicellular ascomycetes. Some species are used in making bread, alcohol, and other foods, and some species are disease-causing, such as the yeast that causes thrush.

A. Obtain a clean glass slide and coverslip. Place a drop of the yeast solution provided on the slide and make a wet mount. Observe under the compound microscope. Are you able to see individual yeast cells?

C. Phylum Basidiomycota

The basidiomycetes get their name from the presence of structures called basidia, on which basidiospores form via meiosis (Figure 43). Unlike ascospores, which are encased inside asci, basidiospores are attached to the tops of basidia (*basidio-* = pedestal). The most familiar basidiomycetes are the mushrooms.

A. Observe any preserved or fresh basidiomycete specimens on display.

B. Obtain a fresh button mushroom. The "gills" of the mushroom are where the basidia with basidiospores are located. Using forceps, gently pluck a gill from a mushroom and make a wet mount of it with a drop of water. Observe the slide under the compound microscope. Are you able to see any basidiospores?

C. Obtain a prepared slide of *Coprinus*. This slide is of a cross section through the cap of the mushroom, with the gills extending into the center. Observe the slide under the compound microscope, focusing on a gill. Draw what you see and label a basidium and basidiospores.

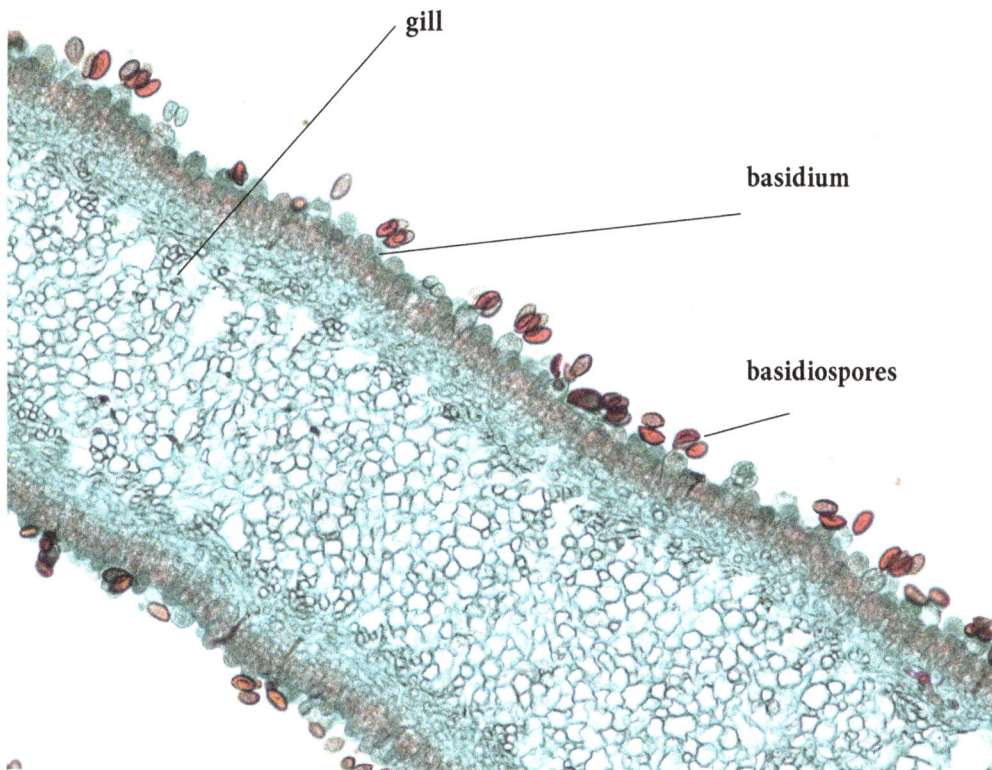

FIGURE 43. Longitudinal section through a gill of *Coprinus* sp., showing basidia with basidiospores, 225X.

D. Lichens

A lichen is an organism consisting of a symbiotic relationship between a green alga or cyanobacterium and an ascomycete (or sometimes basidiomycete) fungus. The photosynthetic activities of the green alga or cyanobacterium provide food for the fungus. Lichens are extremely hardy, surviving in harsh environments and living on a variety of substrates, including rocks, bark, and soil. There are different types of lichens, named according to their body form. Lichens with shrub-like bodies are fruticose lichens, those with leaf-like bodies are foliose lichens, and lichens that encrust their substrate are crustiose lichens. Observe any lichens on display and try to determine whether they are fruticose, foliose, or crustiose.

ADDITIONAL ACTIVITY:

Observe any rotting food on display under a dissecting microscope. Are you able to determine the phylum of fungus growing on the food? In the space below, list each food on display, and for each, the phylum of fungus growing on the food.

QUESTIONS

1. Fungi are saprophytic. What does this mean? Describe in general how a saprophyte feeds.

2. What substance is found in the cell walls of most fungi?

3. How are fungi important economically?

4. Contrast the three main fungal phyla according to their reproductive structures: List the name of the reproductive structure found in in each phylum and the name of the type of spore produced. What process of cell division (mitosis or meiosis) results in the production of spores in each phylum?

5. What are the three main body types seen in lichens? Provide a brief description of each.

6. As mentioned, the fungus of a lichen gets its food from the products of the photosynthetic activities of the alga or cyanobacterium. What is one way in which the alga or cyanobacterium might benefit from this symbiotic relationship?

7. Fill in the table below to contrast visible characteristics of the three fungal phyla seen in lab. How can you distinguish between these phyla, microscopically and macroscopically, on a lab exam?

TABLE 7. A Comparison of Three Major Fungal Phyla.

	Zygomycota	Ascomycota	Basidiomycota
One example organism seen in lab (either the common name or genus)			
At least one visible characteristic of this group (besides color) that helps you distinguish it from other fungal groups (either microscopic or macroscopic)			

ANIMALS

INTRODUCTION

The animals are a large and extremely diverse group of organisms that will be covered over several lab periods. Despite their diversity, all animals share a few common characteristics, including the following: 1) All animals are multicellular eukaryotic heterotrophs; 2) animal cells have no cell walls; 3) most animals ingest their food; 4) animals are capable of movement during all or part of their life; and 5) animals are diploid and produce gametes via meiosis.

The criteria used to classify animals have changed over the years as laboratory techniques have evolved. As with the protists, plants, and fungi, DNA analyses of animals have been extremely valuable in helping scientists 1) discover the degree of relatedness between different animal groups; and 2) reconstruct phylogenetic trees. Prior to the development of molecular techniques, the classification of animals was limited to looking at macroscopic and microscopic differences and similarities between organisms. As the use of DNA sequencing and other methods of molecular analysis increased, the existing classification of animals was either supported, or in many cases, refuted. In many cases, molecular analyses have caused a certain group of animals to be reassigned to a different or new taxonomic category. In others, organisms previously thought of as completely unrelated due to dissimilar morphological characteristics were now being placed into the same taxon due to similarities in DNA sequence. Molecular analyses are not the only method used to classify animals, but their contribution to scientists' understanding of the phylogenies of animals and relatedness of animals is substantial.

The criteria used to classify animals will be discussed in lecture in detail. In lab, we will focus on a few of the main differences between animals that are used to describe them:

1. Type of body symmetry

The body of an animal is either asymmetrical, radially symmetrical, or bilaterally symmetrical. Most poriferans (sponges) are asymmetrical. Cnidarians and ctenophores are examples of radially symmetrical phyla (this manual only covers the cnidarians). Bilaterally symmetrical phyla studied in lab are the platyhelminthes, mollusks, annelids, nematodes, arthropods, echinoderms, and chordates (adult echinoderms possess pentaradial, or five-sided, radial symmetry, but echinoderm larvae are bilaterally symmetrical).

2. Number of embryonic tissue layers present

The embryonic tissue layers are endoderm, mesoderm, and ectoderm. These tissue layers, when present, develop into different adult tissues and organs. Endodermal tissue develops into the gastrointestinal tract lining, mesoderm develops into muscle, the circulatory system, and most internal organs, and ectoderm develops into the nervous system and the skin. The poriferans are the one phylum that lacks embryonic tissues. All other phyla are either diploblastic (possess two embryonic tissue layers) or triploblastic (possess three embryonic tissue layers). The diploblastic phyla have radial symmetry. The diploblastic phylum we will study in lab is Cnidaria. All other animal phyla are triploblastic and exhibit bilateral symmetry.

3. Presence or absence of a body cavity

Within the triploblasts, the presence or absence of a body cavity known as a coelom is another characteristic used to group animals. A coelom is a body cavity lined with mesodermally derived tissue; it is located between the body wall and the outside of the digestive tract. Only one of the triploblastic phyla studied in lab, phylum Platyhelminthes, is acoelomate (lacks a coelom). All other triploblasts studied in lab are coelomates. (Some biologists refer to nematodes as pseudocoelomates because their body cavity is not completely lined with mesodermally derived tissue.) Although lack of a coelom was once considered a primitive trait, the Platyhelminthes are now known to have evolved from coelomate ancestors. Hence, "acoelomate" is a useful anatomical descriptor, but it is not useful for assessing relatedness between groups.

ACTIVITIES

You will likely complete these activities over more than one lab period; follow instructions given by your lab instructor. As you encounter each phylum that follows, it may be useful to list the classification of the members of that phylum as presented in lecture and in lab. In cases where classification of a group tends to differ slightly depending on the textbook being used, general names (for example "arachnids") are used for animal groups in this manual, and it is your responsibility to use your lecture notes, lab notes, and your textbook to help you with the taxa names corresponding to these animals. Questions for all animal labs may be found at the end of this chapter.

Exercise 1. Phylum Porifera

Poriferans (sponges) are mostly asymmetrical animals that lack embryonic, and in most cases adult, tissues. Sponges possess a few different specialized cell types that are used in activities such as feeding and reproduction (gamete production). Choanocytes are cells that sponges use in feeding. Each choanocyte possesses a flagellum that beats and creates a water current that draws microscopic food particles into the sponge. These food particles are then either engulfed by the choanocytes or transferred to other cells that engulf them. Choanocytes are of evolutionary interest, as they are similar in structure to choanoflagellate protists, suggesting a protistan ancestor for animals.

   Sponges are sessile as adults, and there are both marine and freshwater species. Structures called spicules, made of silica or calcium carbonate, give many sponges structural support.
A. Obtain a preserved *Scypha* specimen (also known as *Grantia* sp.) and observe it under a dissecting microscope.

B. Obtain a prepared slide showing either a cross-section (c.s.) or longitudinal section (l.s.) through a *Scypha* specimen. Draw what you see and label the spongocoel, pores, and choanocytes (if visible).

C. Observe any sponge specimens on display.

Exercise 2. Phylum Cnidaria

This phylum includes the jellies, sea anemones, corals, and hydrozoans. Cnidarians are radially symmetrical and are diploblastic, possessing endodermal and ectodermal tissue. In between these tissue layers is a gelatinous substance called mesoglea. Cnidarians do not have a complete digestive tract or gut; instead, they have a gastrovascular cavity with one opening. Many cnidarians have a life cycle that includes two distinct body forms: a sessile, asexually reproducing polyp stage and a free-swimming, sexually reproducing medusa stage. In those cnidarians that exhibit both stages in their life cycle, the diploid medusae produce haploid eggs and sperm via meiosis. These gametes are released into the water. Fertilization is external (occurs in the water column outside of the parent). The resulting diploid zygote develops into a ciliated, swimming larval stage, which eventually settles, attaches to a substrate, and grows into a polyp. The polyp produces new medusae asexually by budding. Cnidarians are so-called because they possess special cells called cnidocytes that contain nematocysts, which are structures that shoot out of the cnidocytes in response to touch (for example, by prey) or noxious chemicals. Nematocysts often contain a toxin to poison the prey, making it easier for the cnidarian to bring the prey item to its mouth and feed.

In lab, we will focus on three classes of cnidarians: Hydrozoa, Scyphozoa, and Anthozoa.

A. Class Hydrozoa

1. *Hydra*: *Hydra* is a freshwater hydrozoan that exhibits only a polyp stage.

    A. Place a live *Hydra* in a Syracuse dish and observe under the dissecting microscope. Place a live *Daphnia* (aquatic arthropod) in the drop of water with the *Hydra* and observe the interaction between the two animals. Is there any evidence that the *Hydra*'s nematocysts in the tentacles have discharged?

    B. Make a wet mount with a new live *Hydra*. Observe the animal under the compound microscope. While observing, add one drop of 10% acetic acid to the edge of the coverslip. Refocus as necessary, and observe carefully for nematocyst discharge. Sometimes closing the iris diaphragm and decreasing the light intensity can help in viewing the nematocysts.

2. *Obelia*: *Obelia* is a marine hydrozoan that exhibits both a polyp and medusa stage. When observing the polyp stage, you will see different types of polyps on one specimen: feeding polyps and reproductive polyps (Figure 44).

    A. Observe the preserved *Obelia* specimens on display.

    B. Observe a prepared slide of *Obelia* under the compound microscope. Draw what you see and label a feeding polyp and a reproductive polyp.

3. *Physalia*: The genus *Physalia* is a group of marine cnidarians that have both polyp and medusa stages. *Physalia* is a highly specialized colonial hydrozoan, meaning it is a group of polyps and medusae living together to form one individual organism. There are feeding polyps, stinging polyps, and some of the attached medusae house the gametes.

    A. Observe preserved *Physalia* specimens on display.

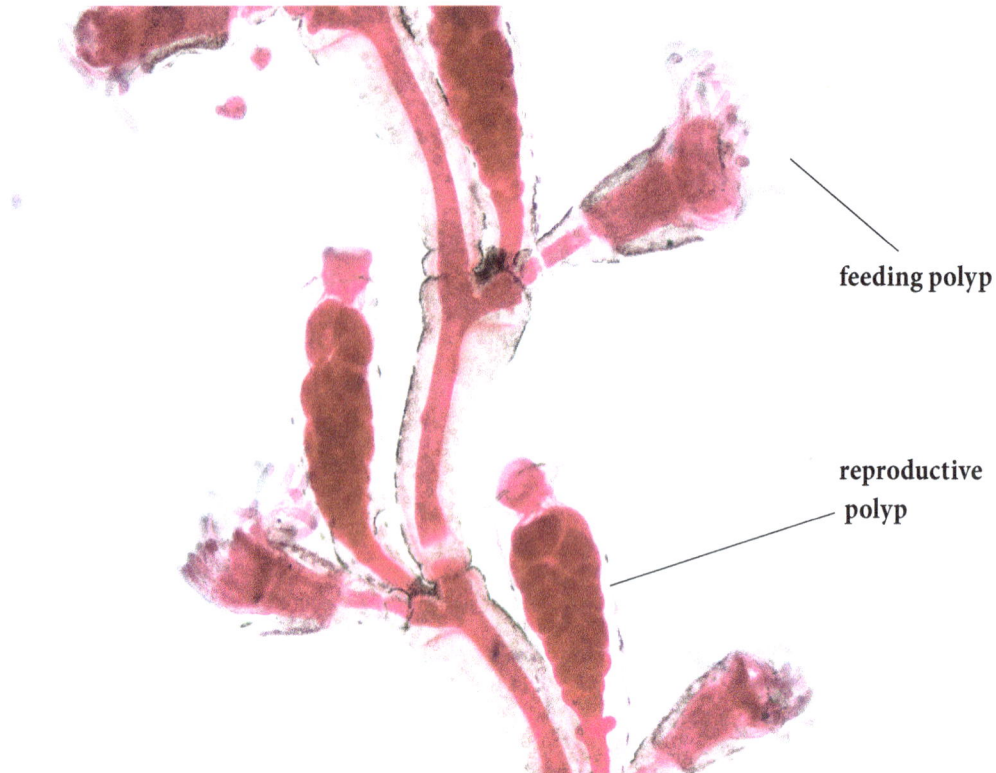

FIGURE 44. *Obelia* sp., w.m., 47x.

B. Class Scyphozoa

Scyphozoans, the jellies, exhibit both a medusa and polyp phase. As described in the introduction, the medusa reproduces sexually, producing gametes via meiosis. After fertilization occurs, the diploid zygote develops into a ciliated larval stage, which then develops into the polyp stage. The tiny polyp is able to reproduce asexually, making more medusae via budding. The medusa phase is the most conspicuous phase in this group.

    1.   Observe preserved scyphozoans on display.

C. Class Anthozoa

Anthozoans include sea anemones and corals, and usually only exhibit a polyp stage.

    1.   Observe preserved and live (if available) sea anemones.

Exercise 3. Phylum Platyhelminthes

Members of this phylum are informally called flatworms. These worms are bilaterally symmetrical, acoelomate triploblasts. Some platyhelminthes possess a gastrovascular cavity with one opening, while some—namely the tapeworms—do not have any type of digestive tract and absorb nutrients through their body wall. Most flatworms are hermaphrodites, but nonetheless successful reproduction in many species requires the exchange of egg and sperm between two individuals.

    In lab we will focus on three classes of Platyhelminthes: Turbellaria, Trematoda, and Cestoda.

A. Class Turbellaria

Turbellarians are the free-living flatworms, living mostly in freshwater and marine environments. One of the more common members of the class is the genus *Dugesia*, or *Planaria*. In fact, turbellarians are sometimes informally called planarians (Figure 45).

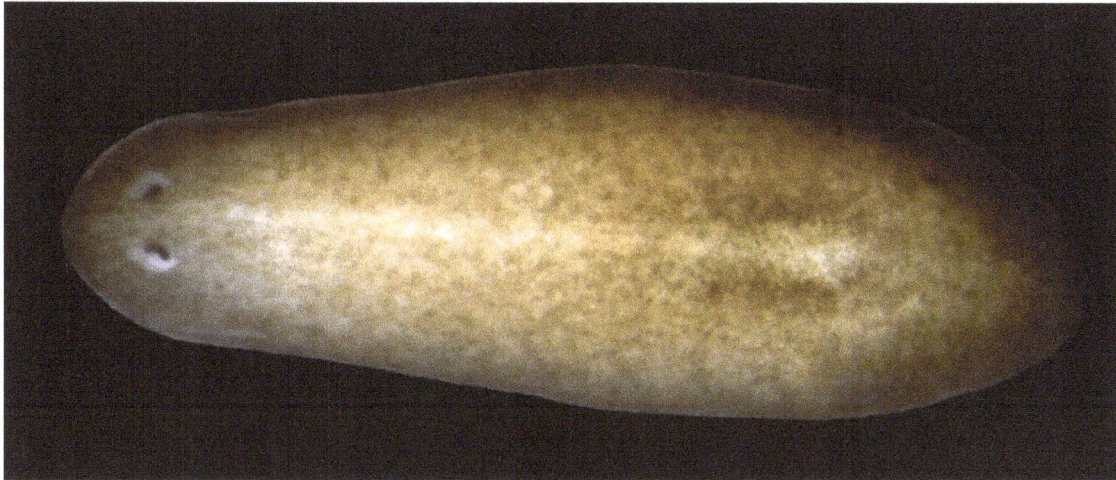

FIGURE 45. *Schmidtea* sp., a planarian.

1. Obtain a live planarian from the designated area in the lab: Use a transfer pipette to gently place the animal into a Syracuse dish. Take the dish back to your lab bench and observe the planarian under a dissecting microscope, without turning on the lamp.
2. The eyespots are clusters of pigmented cells located near photosensitive receptors (Sherman and Sherman, 1976). Shine a penlight on the planarian. How does the planarian react to light?
3. Place a small piece of liver or cooked egg yolk in the dish with the animal and observe its response. Does the animal move toward the liver or egg yolk? Are you able to observe any feeding?
   When you are finished observing the planarian, return it to the area designated in the lab, and rinse and return the Syracuse dish.
4. Obtain a prepared slide of a whole mount of a planarian and observe under the compound microscope. Draw what you see and label the eyespots and the pharynx.

B. Class Trematoda

Trematodes, the flukes, are all parasitic. Their life cycles are complex, and they live in a variety of hosts, depending on the particular species.

1. Observe preserved specimens on display.
2. Obtain a prepared slide of a whole mount of *Clonorchis sinensis*, the Chinese liver fluke (Figure 46). Draw what you see and label the following: sucker, pharynx, intestine, uterus, ovary, testis.

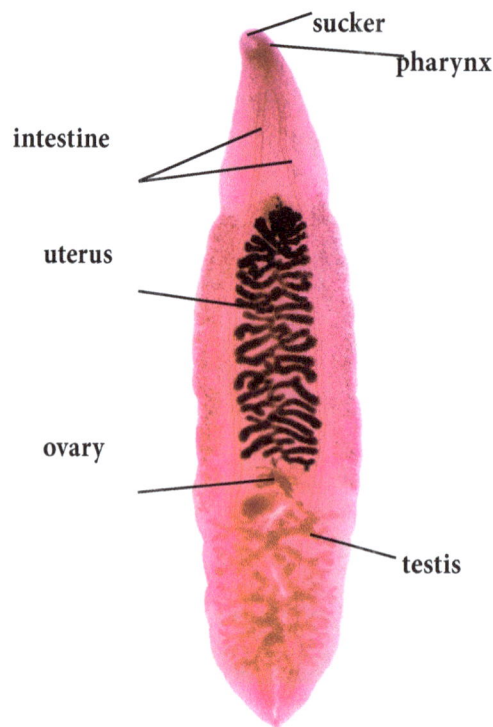

FIGURE 46. *Clonorchis sinensis,* **the Chinese liver fluke.**

C. Class Cestoda

NOTE: Some textbooks use Cestoidea or other class names for this group.

The cestodes are the tapeworms. They are parasitic, and as with the trematodes, they live in a variety of hosts and have complex life cycles. The cestode body is divided into segments called proglottids. Proglottids are not true body segments as seen, for example, in the annelids. Each proglottid contains a set of reproductive organs, both ovaries and testes (Figure 47). Tapeworms do not possess a digestive tract; they absorb their nutrients from the digestive tract of their host via diffusion across their body wall. Their extensive reproductive system, their lack of digestive tract, along with their hooks and suckers, make tapeworms well adapted for parasitism.

1. Observe preserved specimens on display.
2. Obtain a prepared slide of *Taenia pisiformis*, a parasite of dogs, and observe under the compound microscope. Note that these slides often have four different preparations: a region showing the scolex at the anterior end, immature proglottids, mature proglottids, and gravid proglottids. Be sure to observe the slide at a few different magnifications to see as many structures as possible. Draw what you see and label the following: scolex, hooks, suckers, uterus, ovary, testes.

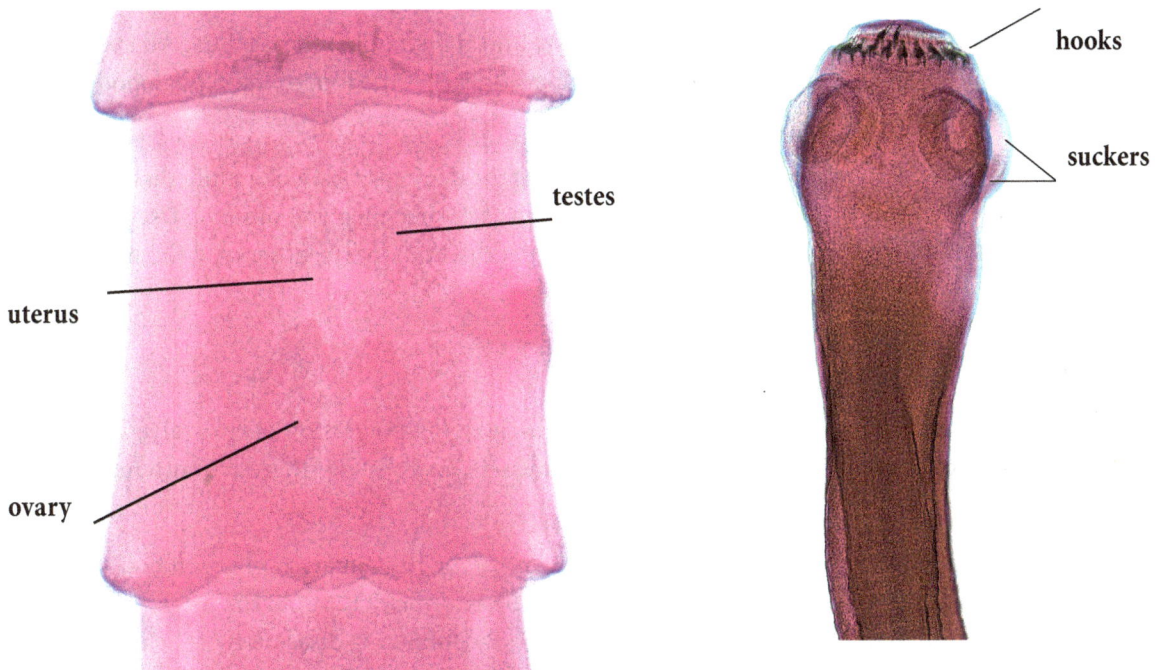

FIGURE 47. *Taenia pisiformes,* **mature proglottid, 685x (left) and scolex, 22x (right).**

Exercise 4. Phylum Mollusca

Mollusks are bilaterally symmetrical triploblasts. They are the first animals mentioned so far to possess a coelom, although the coelom in most mollusks is reduced to the pericardial sac surrounding the heart. Mollusks are also the first animals mentioned so far to have a complete digestive tract (it has two openings). Most mollusks possess an open circulatory system, although the cephalopods have a closed circulatory system. The body of a mollusk has three main parts: 1) A muscular foot on the ventral side of the animal, functioning in locomotion; 2) a visceral mass that houses most of the internal organs, including the respiratory and excretory organs and the heart; and 3) the mantle, which is a tissue that encloses the visceral mass; the mantle secretes the shell in those mollusks that have one. The space between the visceral mass and the mantle is the mantle cavity, and in some molluscan classes, respiratory structures such as gills extend into the mantle cavity.

In lab we will focus on three molluscan classes: Gastropoda, Cephalopoda, and Bivalvia.

A. Class Gastropoda

This class includes snails, slugs, abalone, limpets, and nudibranchs. There are terrestrial, marine, and freshwater gastropods. Most possess a protective shell, but this is lost in the slugs and nudibranchs, for example.

1. Observe preserved and live (if available) specimens on display. Try to identify the three main molluscan body parts in each specimen you observe. Use any pictures available to help you in your identification of the body parts, especially the mantle and visceral mass, which may not be visible without the aid of a diagram or picture.

B. Class Cephalopoda

The cephalopods include squids, octopuses, cuttlefish, and the chambered nautilus. All species are marine. The molluscan body plan is present in cephalopods, except the foot of the cephalopod body is modified into eight arms.

1. Observe preserved specimens on display, including the demonstration of a squid dissection. Try to identify the three main molluscan body parts in each specimen you observe. Use pictures in your textbook to help you in your identification of the body parts, especially the mantle and visceral mass.

C. Class Bivalvia

The bivalves are named after a distinctive feature of this class, the possession of two valves, or shells. Familiar members of class Bivalvia include the clams, mussels, and oysters. There are both marine and freshwater species.

1. Observe preserved bivalves on display. Try to identify the three main molluscan body parts in each specimen you observe. These features may be more difficult to see in the preserved bivalves on demonstration due to the presence of the valves; the following dissection of the clam will be of great help in locating the mantle, visceral mass, and foot, as well as other structures.

A note about dissections:

Since this is the first animal to be dissected this semester, it is important to first understand some basic rules of and techniques for any dissection:

1. Always wear gloves when dissecting preserved or fresh specimens.
2. Different specimens often require the use of different types of tools for dissection. For example, not all specimens are easily opened up with a scalpel: some require using scissors, while others should only be dissected with a pin due to the fragility of the tissue. Pay attention to any announcements and read directions carefully before beginning to dissect a specimen. Make sure you understand the terms **anterior**, **posterior**, **dorsal**, and **ventral**.
3. Work carefully and methodically in order to best visualize the parts of the specimen.
4. Be careful and use common sense when working with sharp dissecting tools! Always cut away from you when using scissors or a scalpel.
5. When you are finished with your dissection, place the specimen and any tissue parts remaining in the tray in the waste area designated by your instructor. Be sure to remove any pins from the tray before rinsing. Rinse the tray thoroughly and leave it to dry.

Dissection of the freshwater clam

1. Put on gloves. Obtain a dissecting tray and the following dissecting tools: scissors, pins, blunt probe, and a sharp probe.

2. Before opening the clam, identify the dorsal and ventral sides of the animal, as well as the anterior and posterior ends. The dorsal side of the clam is the side where the two valves attach, and the ventral side is the side where the valves open. Near the point of valve attachment on the dorsal side, you will find an area called the **umbo** on each valve. The umbo is the place where the valve began to develop. The anterior end of the clam is often, but not always, the end closest to the umbo.

3. To open the clam, you must use scissors to cut the posterior and anterior adductor muscles, which are the two muscles that keep the valves closed. First, carefully insert the scissors in between the valves at the posterior end of the clam. By moving the scissors, you should be able to detect the area of the **posterior adductor muscle**. Cut the muscle so that the valves are loosened. After cutting this muscle, use your hands to carefully pry the valves apart. If you are able to open the valves, move to Step #4. If the valves will not separate, use your scissors to cut the anterior adductor muscle, using the same technique you used to cut the posterior adductor muscle. Ask your lab instructor for assistance, if necessary.

4. Once you are able to separate the valves, do so carefully and notice the brownish tissue adhering to the inside of each valve. This tissue is the **mantle**, and the space surrounded by the mantle tissue is the **mantle cavity**. The mantle secretes the shell in those species that possess one. The rim of the mantle is usually darker in color, and at the posterior end, this rim forms two **siphons**, one for water intake, and one for water expulsion. When you separated the valves, the siphons were broken, but try to find the areas of the mantle rim that formed the siphons.

5. Next, locate the **gills**, **foot**, **visceral mass**, **posterior adductor muscle (now cut)**, **mouth**, and **anus**. The mouth is most easily located by looking for the **labial palps** that surround it. The palps aid in guiding food into the clam's mouth.

6. Locate the **heart**. The heart is very delicate and is located on the dorsal side of the animal, almost midway between the anterior and posterior ends. It is surrounded by a pericardial sac (the space between this sac and the heart is the coelom in these animals). To most successfully find the heart, look directly down on the dorsal side of the clam, and, using a dissecting pin or sharp probe, carefully pick away the sac to expose the heart. Use pictures and diagrams to help. The heart has perforations in it; recall that bivalves and most mollusks have an open circulatory system.

7. Using a scalpel, carefully make a longitudinal slice through the visceral mass. Try to locate the **intestine** running through the mass. The intestine connects anteriorly and dorsally to the stomach (which is often difficult to locate). Darker tissue near the anterior end of the visceral mass is the **digestive gland**, which has both digestive and detoxification functions.

8. Before discarding your clam, make sure you are able to identify the structures listed in the chart that follows. Refer to Appendix 2 for assistance with identifying the structures. When finished, discard your clam and any tissue parts in the hazardous waste receptacle designated by your instructor. Remove any pins from the tray, rinse the tray, and leave it to dry.

Structures and areas to identify in the clam:

Structure	Function
anterior and posterior ends	N/A
dorsal and ventral sides	N/A
valve	
umbo	N/A
mantle	
mantle cavity	N/A
siphons	
gills	
foot	
visceral mass	
posterior adductor muscle	
mouth	
labial palps	
anus	
heart	
intestine	
digestive gland	

Exercise 5. Phylum Annelida

The annelids are the segmented worms. Like the mollusks, they are bilaterally symmetrical triploblasts, and they have a complete digestive tract. Their coelom is segmented and serves as a hydrostatic skeleton: The muscles of the body wall contract against the pressure exerted by the fluid in the coelom, and this results in movement of the animal's body. Unlike most mollusks, annelids have a closed circulatory system. They are hermaphroditic, possessing both male and female reproductive organs, but they cannot self-fertilize.

The three main groups of annelids are the oligochaetes, polychaetes, and hirudineans. These taxa are referred to as classes in some sources. Furthermore, the oligochaetes and hirudineans are often categorized as the clitellates due to their possession of a reproductive structure called a clitellum.

A. Oligochaetes

Earthworms are probably the most familiar oligochaetes. There are other terrestrial oligochaetes as well, in addition to aquatic species. The oligochaetes get their name from the chaetae (= setae) found on each segment. Chaetae are small, hairlike structures used for locomotion. They are difficult to see with the naked eye, but can be felt as a bristly line running along the sides of the worm's body. Oligochaetes have fewer chaetae (*oligo-* = few) polychaetes (*poly-* = many).

1. Observe any preserved oligochaetes on display.
2. Obtain a live earthworm and place it on a paper plate. Wash your hands before handling the earthworm. Bring the worm back to your bench, and with bare hands, run your fingers along the worm's body. Can you detect the chaetae? Distinguish between the anterior and posterior ends and the dorsal and ventral sides of the worm. Locate the clitellum, which forms mucus during mating and also forms a cocoon in which fertilization occurs (Figure 48).

FIGURE 48. *Lumbricus terrestris,* **the earthworm. The clitellum is the light-colored structure near the anterior end of the worm.**

When you are finished observing the earthworm, place it back in the container in the laboratory and wash your hands.

3. *Lumbriculus* sp. is a tiny freshwater oligochaete. Because of its small size and translucent cuticle, it is a good specimen to use for observing the activities of the annelid's closed circulatory system. Using a transfer pipette, obtain some live *Lumbriculus* worms and place them into a Syracuse dish for observation. Take the dish back to your laboratory bench, and observe the worms under a dissecting microscope. Patient and careful observation will allow you to see blood circulating through the vessels of the worm's circulatory system. When finished, carefully transfer the worms back into their original container.

DISSECTION OF THE EARTHWORM

You may choose to dissect either a live or preserved earthworm. If you dissect a live earthworm, obtain a worm and place it in a beaker containing 7–10% ethanol. Once the worm appears narcotized, proceed with the instructions for dissection that follow.

1. Put on gloves. Obtain a dissecting tray, dissecting scissors, and pins. Pin the worm onto the tray, ventral side facing down, by placing one pin into the posterior end and another as far anterior as possible.

2. Using the pointed end of scissors or a pin, make a cut into the dorsal side of the worm near the posterior end. Slit the worm on the dorsal side from posterior to anterior ends. Pin the body wall of the worm back on either side, beginning at the anterior end and ending about halfway down the worm's body. In the anterior-most portion of the worm, you will find a tiny pair of **cerebral ganglia**, which form the brain of the animal. The **pharynx**, **crop**, **gizzard**, and **intestine** are all parts of the digestive tract. The muscular pharynx allows the animal to take in dirt/food matter, while the crop and gizzard serve in storing and grinding food, respectively. The intestine allows for digestion of and passage of food through the body. The **seminal vesicles** store sperm prior to mating, and **seminal receptacles** receive sperm during mating. These structures are white in color and are located anterior of the clitellum. Before discarding your earthworm, be sure you are

able to identify the structures listed in the chart that follows. Refer to Appendix 2 for assistance with identifying the structures.

Structures and areas to identify in the earthworm:

Structure	Function
brain (cerebral ganglia)	
pharynx	
crop	
gizzard	
intestine	
coelom	
clitellum	
seminal vesicles and receptacles	
paired "hearts"	

B. Hirudineans

Hirudineans are the leeches. Members of this class are mainly predators and scavengers. Some are ectoparasites, attaching themselves to their host using anterior and posterior suckers and feeding on the blood of the host. Leeches live in freshwater environments such as streams and ponds; a few species are marine.

1. Observe any preserved and live specimens available, and in the space below, make any notes or drawings that will help you study for quiz and practical questions on the leeches. What *visible* features distinguish these animals as annelids? As hirudineans?

FIGURE 49. **Leech, dorsal view.**

C. Polychaetes

The polychaetes are mostly marine. Their name, as previously mentioned, means "many chaetae." The chaetae of polychaetes are located at the tips of parapodia, protruding structures located on each segment. Parapodia function in locomotion and in gas exchange in these worms.

1. Observe any preserved or live specimens on display. Identify the parapodia and chaetae. Refer to Figure 50.

   Use the space below to make any notes or drawings that will help you study for quiz and practical questions comparing the oligochaetes and polychaetes. What *visible* features distinguish these animals as annelids? As oligochaetes or polychaetes?

FIGURE 50. *Nereis sp.*, a polychaete.

Exercise 6. Phylum Arthropoda

This phylum is extremely diverse, with members living in terrestrial, marine, and freshwater environments and members capable of flight. Phylum Arthropoda also contains the greatest number of animal species, due to the most numerous class of arthropods, the insects. Despite their diversity, all arthropods share certain characteristics. Arthropods possess paired, jointed appendages, and it is for this characteristic they are named (*arthropod* = jointed foot). In addition, arthropods possess a hard exoskeleton containing chitin (the exoskeleton of crustaceans also contains calcium carbonate). The exoskeleton serves as a site for muscle attachment and also protects the animal from predation and desiccation (terrestrial species). One disadvantage to having this exoskeleton is that in order to grow, the animal must shed, or molt, the exoskeleton and grow a new one. This molting process is called ecdysis. During ecdysis, the animal is vulnerable to predation and must hide. Other characteristics of arthropods include an open circulatory system and body segmentation. Neither of these features is unique to arthropods. However, the segmentation seen in arthropods is more specialized than that seen in annelids: Segments in arthropods are often adapted for specific functions, for example, movement and sensing the environment.

The classification of arthropods is less straightforward than that within other animal phyla mentioned thus far. The particular taxa used by biologists (and in textbooks) are fairly consistent (for example "Crustacea," "Hexapoda"), but whether a particular taxon is considered a subphylum or class is not. For example, some biologists consider "Crustacea" to be a class, while others consider this taxon to be a subphylum. These differing views mean that textbook classifications often differ as well. The following description of arthropod groups omits any reference to "class" or "subphylum." Instead, just the generally agreed-upon taxa names are used to describe the groups. It is your responsibility to note whether a particular group is a class or subphylum, according to what you have learned in lecture and from your textbook.

A. Crustaceans

Many crustaceans, including crabs, lobsters, shrimp, and barnacles, are aquatic. Isopods ("pill bugs") are terrestrial crustaceans. Distinguishing characteristics of crustaceans include two pairs of antennae and also biramous (branched) appendages.

1. Look at any preserved and live specimens available, and in the space below, make any notes or drawings that will help you study for quiz and practical questions on the crustaceans. What *visible* features distinguish these animals as arthropods? As crustaceans?

Dissection of the crayfish

1. Put on gloves. Obtain a preserved crayfish, dissecting tray, scissors, pins, and a sharp or blunt probe.
2. Before dissecting your crayfish, identify the following external features and indicate their functions, where appropriate, in the table that follows this description: **cephalothorax** (fused head-thorax region), **carapace** (another term for the exoskeleton covering the cephalothorax region), **abdomen**, **tail**, **antennae**, **mandibles**, **chelipeds**, **walking legs**, and **swimmerets**. Determine whether your crayfish is male or female by looking on the animal's ventral side. Male crayfish have a pair of swimmerets that are modified for copulation located near where the cephalothorax and abdomen meet. Appendix 2 will be helpful as you attempt to locate these organs.
3. Hold the crayfish so that you are looking at one side of the animal and locate the border between the head region and thorax region. Using scissors, begin cutting along this border until the cut is complete, i.e., it runs up one side and down the other. Next, cut along the posterior border of the carapace, between the thorax and abdomen, until the cut is complete. Gently pry back the carapace and use a probe to peel away any tissue that is adhering to the carapace. Work carefully, as the heart is in this region and is delicate. Locate the feathery **gills** running along both sides of the animal. These function in respiration. The **heart** is located on the dorsal side of the animal, about midway between where the anterior and posterior borders of the carapace were. Use Appendix 2 and any demonstrations on display in the laboratory to help you locate it. Note the small perforations in the heart, illustrating the open circulatory system of these animals.
4. Once you have located the heart, gently remove it and the gills using either a probe or scissors. Just ventral to the heart lie the **gonads** of the crayfish. In preserved specimens, ovaries and testes are indistinguishable unless a female is gravid (eggs are present). Ventral and slightly lateral to the gonads are the **digestive glands, which secrete digestive enzymes.** These often extend anteriorly. Drawings showing a lateral view or a cross-sectional view of the crayfish are extremely helpful

as you attempt to locate these organs. The **stomach** is anterior to the heart and is near the dorsal surface. The **esophagus** extends anteriorly from the stomach into the head region.

5. Using scissors, make a triangular cut in the head area of the exoskeleton. Locate the circular, paired **green glands** in the head. These organs function in osmoregulation (electrolyte balance). In preserved specimens, these are often not green; they are light to dark brown.

6. Once again, locate the esophagus. Note the circumesophageal nerve cords on either side of the esophagus. These meet anteriorly at the **cerebral ganglion ("brain")**.

7. Before discarding your crayfish, be sure you are able to identify the structures in the chart below. Refer to Appendix 2 for assistance with identifying the structures.

Structures and areas to identify in the crayfish:

Structure	Function
cephalothorax	
carapace	
abdomen	
tail	
antennae	
mandibles	
walking legs	
chelipeds	
swimmerets	
gills	
heart	
gonads	
digestive glands	
stomach	
esophagus	
green glands	
cerebral ganglion	

B. Insects

The insects are the most numerous group of arthropods. Insects have traditionally been classified as members of class Insecta and of subphylum Hexapoda; however, recent evidence has led some scientists to place them within the Crustacea.

1. Look at any preserved and live specimens available, and in the space below, make any notes or drawings that will help you study for quiz and practical questions on the insects. What *visible* features distinguish these animals as arthropods? As insects? (Hint: How many legs do all of the insect specimens possess?)

Dissection of the grasshopper

1. Put on gloves. Obtain a preserved grasshopper, dissecting tray, scissors, and a sharp or blunt probe.
2. Before cutting open the grasshopper, view its mouthparts. Try to locate the **mandibles**, paired jaw-like appendages near the mouth. Using the scissors, carefully remove the exoskeleton so that you are able to view the internal structures from the dorsal side.
3. Distinguishing between different structures in the preserved grasshopper can be challenging due to the fact that all structures are similar in color. Use diagrams and pictures to help you identify the **esophagus**, **crop**, **stomach**, **intestine**, **gastric caecae**, and **Malpighian tubules**. The gastric caecae are finger-like projections originating at the intersection of the crop with the stomach; they secrete digestive enzymes into the stomach. Malpighian tubules are much smaller, hairlike projections located posterior to the gastric caecae. Malpighian tubules have an osmoregulatory function.
4. Before discarding your grasshopper, be sure you are able to identify the structures listed in the chart below. Refer to Appendix 2 for assistance with identifying the structures.

Structures to identify in the grasshopper:

Structure	Function
mandibles	
esophagus	
crop	
stomach	
intestine	
gastric caecae	
Malpighian tubules	

C. Myriapods

The myriapods include the millipedes and the centipedes. Millipedes differ from centipedes in the number of walking legs found on each segment of the animal. Millipedes possess two legs per segment, while centipedes have one pair of legs per segment.

1. Observe any preserved and live specimens available. What *visible* features distinguish these animals as arthropods? As millipedes or centipedes?

D. Chelicerates

All of the arthropods observed thus far have antennae as their first appendages and jaw-like mandibles as mouth parts (the so-called "mandibulates"). The chelicerates are named after their possession of pincer-like appendages called chelicerae. The chelicerae are the first pair of appendages of the chelicerates, and in some specimens (arachnids), the chelicerae possess hollow, venomous fangs. Chelicerates lack antennae. The chelicerates include horseshoe crabs, sea spiders, and the arachnids. It is this last group that we will focus on in lab. The arachnids include spiders, scorpions, mites, and ticks.

1. Observe any preserved and live specimens available. What *visible* features distinguish these animals as arthropods? As arachnids? How many pairs of legs do arachnids possess?

Exercise 7. Phylum Nematoda

The nematodes are the roundworms, and are perhaps the most numerous of all animals. They are mostly free-living, but include many important parasites of plants, animals, and humans. They are pseudocoelomate, with their pseudocoelom serving as a hydrostatic skeleton. The body is covered with a flexible cuticle that must be molted in order for the worm to grow. The representative specimen we will focus on in lab is the parasitic roundworm *Ascaris* sp. (Your lab instructor will inform you if other, free-living species are available for observation.) Details about the life cycle of *Ascaris* may be found in your lecture notes and textbook.

1. Observe any preserved *Ascaris* specimens available. What *visible* features distinguish the male *Ascaris* specimens from the female *Ascaris* specimens? What *visible* features distinguish the roundworms from the flatworms (Platyhelminthes) and segmented worms (annelids)?

2. If live free-living nematodes are available, observe their eel-like movement, a result of possession of only longitudinal muscles in the body wall.

3. Your lab instructor may choose to have you dissect a preserved *Ascaris* specimen. If so, more information will be provided in lab.

Exercise 8. Phylum Echinodermata

The echinoderms include sea stars, sea cucumbers, sea urchins, and sand dollars. Larval echinoderms live in the water column and exhibit bilateral symmetry, while the sessile adults possess radial symmetry. The radial symmetry seen in echinoderms is pentaradial, meaning it is arranged around five planes of symmetry.

Echinoderms are named after a key feature of this phylum: their spiny endoskeleton (*echino-* = spiny; *derm-* = skin). Another unique characteristic of this phylum is a water vascular system. The water vascular system is a set of tubes/structures inside the animal that are filled with mainly seawater. It functions in locomotion and gas exchange. Consult your lecture notes and textbook for more information on the water vascular system.

A. Class Asteroidea

Members of class Asteroidea, the sea stars, are probably the most familiar echinoderms, living in a wide range of marine environments, ranging from the ocean bottom to the intertidal zone. The body of a sea star exhibits the classic echinoderm pentaradial symmetry described above.

1. Observe any preserved or live specimens available. What *visible* features distinguish these animals as echinoderms?

B. Class Holothuroidea

Holothuroideans are the sea cucumbers, named after their body shape. Like the sea stars, sea cucumbers are found in many marine environments figure 51.

FIGURE 51. *Sclerodactyla* sp., a sea cucumber.

1. Observe any preserved or live specimens available. What *visible* structures on these specimens could help you identify this animal as an echinoderm?

C. Class Echinoidea

The echinoids include sea urchins and sand dollars. The spiny endoskeleton of these echinoderms is especially visible in live sea urchins, whose bodies possess sharp spines. These spines fall off when the animal dies, leaving a hollow "test" (sand dollars also leave behind tests when they die). Sea urchins and sand dollars, like other echinoderms, possess tube feet, although these are not as conspicuous as they are in sea stars or sea cucumbers and can be difficult to see.

1. Observe any preserved or live specimens available.

Exercise 9. Phylum Chordata

A diverse phylum, the chordates include animals ranging from sea squirts to humans. Despite this diversity, there are four unifying characteristics that all chordates possess during all or at least part of their lives. These hallmark chordate features are

1. A **dorsal, hollow nerve cord** (this becomes the spinal cord in vertebrates)
2. A flexible rod called a **notochord** that extends the length of the body (this becomes part of the vertebral column in vertebrates); the notochord is ventral to the nerve cord
3. **Pharyngeal gill slits**
4. A **postanal tail**

Most chordates fall into one of three subphyla: Urochordata, Cephalochordata, or Vertebrata. As you complete the exercises that follow, use your lecture notes and textbook to further help you understand both the unifying features of chordates and the features that distinguish the subphyla from one another.

A. Subphylum Urochordata

Urochordates are the tunicates, which comprise the sea squirts and salps. Tunicates are found in both marine and freshwater environments. Larval tunicates swim in the water column. Adult sea squirts are sessile (Figure 52), attaching to rocks or other substrates in the ocean or other body of water, while adult

FIGURE 52. An adult sea squirt.

salps are free-swimming. At first glance, it may be difficult to see why these animals are chordates, but upon closer observation, certain characteristic chordate features may be seen. Larval tunicates possess all four chordate characteristics.

1. Observe any preserved adult tunicates available. Adult tunicates do not possess all four chordate characteristics. What chordate feature(s) is/are visible in the preserved specimens?

B. Subphylum Cephalochordata

Cephalochordates are commonly called lancelets, or amphioxus. Throughout their lives, lancelets possess all four chordate characteristics. These animals live with their tails buried in sand at the bottom Figures 53 & 54 of the ocean.

1. Obtain a preserved lancelet and observe it in a Syracuse dish under a dissecting microscope. Which of the four chordate characteristics are visible?
2. Under the compound microscope, view a prepared slide of a whole mount of a lancelet. Identify the postanal tail, pharyngeal gill slits, notochord, and dorsal hollow nerve cord.
3. Under the compound microscope, view a prepared slide of a cross section through the pharyngeal region of a lancelet. Draw what you see and label the dorsal hollow nerve cord, the notochord, the pharyngeal slits, the postanal tail, the coelom, and the myotomes.

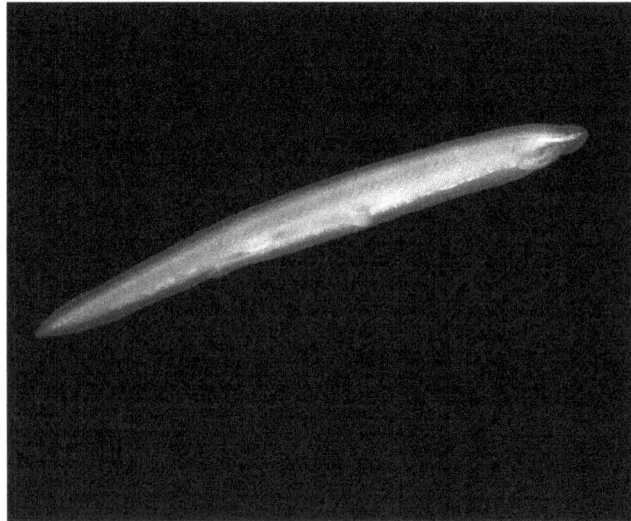

FIGURE 53. *Branchiostoma* sp., a lancelet.

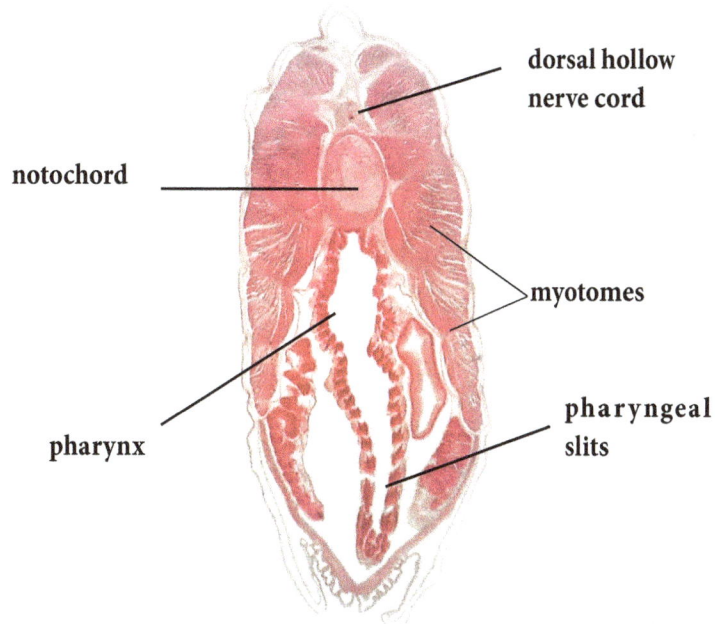

FIGURE 54. Lancelet, pharyngeal region c.s.

C. Subphylum Vertebrata

The vertebrates, as their name suggests, are the chordates that possess a vertebral column surrounding the dorsal nerve cord (spinal cord). Vertebrate taxonomy changes as scientists develop further understanding of the phylogeny of these animals. Similar to many of the other organisms discussed thus far, classification of vertebrates will differ between textbooks as research progresses and scientists debate their taxonomy. For vertebrate classes that have not changed much in their name or composition recently, the class names are given below. For those groups whose classification changes more frequently, common names are listed and you will need to fill in the class name where indicated. Remember to consult your lecture and lab notes and textbook.

1. Lampreys and hagfish

Lampreys and hagfish are the jawless vertebrates. Other than this feature and their eel-like body form, they are different from each other in several ways. Record the classes of each of these used by your text below:

Lamprey = class_____

Hagfish = class_____

Lampreys live in marine and in freshwater environments. They are predators and as adults they feed mainly on the body fluids of living fish. To do this, they attach themselves to the body of a living fish using the sharp teeth found in their circular mouths. Hagfish are marine, living at and feeding at the bottom of the ocean. Their diet consists mostly of dead organic matter that has sunk to the bottom. Although often classified as vertebrates, they actually have no vertebral column. Consult your lecture notes and textbook for more information on lampreys and hagfish.

A. Observe the preserved lamprey and hagfish specimens available. What are two *visible* features that can help you distinguish between lampreys and hagfish?

For the remaining chordate classes, observe the specimens available and list some unifying features of members of each class in the spaces provided.

2. Class Chondrichthyes

Members of this class are the sharks, skates, and rays. They are the cartilaginous fishes (*chondr-* = cartilage), referring to the fact that their skeletons are made of cartilage. Almost all species are marine, but a few species occupy freshwater habitats. A few visible characteristics distinguish the chondrichthyes from the ray-finned fishes, including visible gill slits and asymmetrical tails.

This class exhibits three main modes of reproduction: oviparity, ovoviviparity, and viviparity. Oviparous species lay eggs, and the young develop in eggs outside the mother and receive nourishment from an attached egg sac. In ovoviviparous species, the young develop inside the mother, but are still nourished via an attached egg sac. Viviparous species give live birth to young, which have been nourished via a placental attachment during development. These modes of reproduction will be discussed further in the reproductive system lab.

Unifying features of chondrichthyans:

3. Ray-finned fishes

The class name for this group often differs between texts. Enter the class name used by your text here:

class_____

The ray-finned fishes are given this name because of the ray-like arrangement of supportive rods in their fins. Ray-finned fishes have skeletons made of bone, and their gills are covered by a bony structure called an operculum.

Unifying features of ray-finned fishes:

After observing the cartilaginous fishes and the ray-finned fishes in lab, list at least two visible characteristics that could help you distinguish between these two fish on a lab practical (even if you already know how to tell the difference between a shark and a perch, list two visible differences between these two!):

4. Class Amphibia

Amphibians include frogs, toads, newts, and salamanders. This class is tied to a watery environment for several reasons: 1) mating occurs in water; 2) eggs are laid in water; 3) larval amphibians are aquatic; and 4) adult amphibians need to live in a moist environment to prevent desiccation of their skin.

Unifying features of amphibians:

5. Class Reptilia

Reptiles include snakes, lizards, turtles, and birds. This class, unlike previous classes mentioned, is completely adapted for life on land. There are several characteristics that make reptiles adapted to terrestrial environments, and one of the most important is the amniotic egg. This egg allows the embryo inside to safely develop in a terrestrial environment. It protects the embryo from desiccation, allows for sufficient gas exchange, and nourishes the embryo. Refer to any diagrams or pictures available in lab of the amniotic egg.

What is another feature of reptiles that makes them completely adapted for a terrestrial environment?

Unifying features of reptiles:

6. Class Mammalia

Mammals are named after a feature unique to this class: mammary glands with which to nourish their young. Another unifying characteristic of this class is the presence of fur or hair. A majority of mammals are viviparous and have a well-developed placenta, but one mammalian group lays eggs (the monotremes). Another viviparous group of mammals nourishes young via a placenta, but the young are born early and development finishes within a pouch on the mother (marsupials). In those mammals possessing a placenta during pregnancy, an amniotic sac encloses the embryo. More information on this will be provided in the reproductive system lab.

Unifying features of mammals:

Exercise 10. Mammalian Dissection: The Laboratory Rat

In today's lab you will begin your dissection of the laboratory rat with an examination of its external features. The organ systems of the rat will be dissected and examined in subsequent lab periods.

External features of the rat

1. Put on gloves. Obtain a preserved rat and dissecting tray. No dissecting instruments are needed.
2. Determine whether your specimen is male or female. In male rats, a **scrotum** containing the **testes** is present. The **penis** of male rats is only clearly visible during copulation, although you should be able to locate the **urethral opening**. If your rat is female, identify the **urethral opening** and the **vaginal opening**.

Identify the following structures and areas:

head	nostrils
thorax	posterior and anterior ends
abdomen	dorsal and ventral sides
mouth/oral cavity	anus

When finished, seal your rat in a plastic bag and label the bag with your names and section number. Store in the location indicated by your lab instructor.

QUESTIONS

1. In what phylum are choanocytes found? What is the function of choanocytes? What clue do they provide about the ancestor of animals?

2. State the two body types seen in the cnidarian life cycle. Which body type carries out asexual reproduction? Which body type carries out sexual reproduction? State which type(s) is/are found in each of the three cnidarian classes observed in lab.

3. You are given a piece of an organism and told it is either poriferan or cnidarian. You place the sample underneath the compound microscope, looking for key cell types found in these phyla.

   A. What is a cell type unique to sponges that could help you identify the specimen as poriferan?

   B. What is a cell type unique to cnidarians that could help you identify the specimen as cnidarian?

   C. While looking at the sample under the microscope, you see two distinct tissue layers present. Does this information help you identify the phylum (Porifera or Cnidaria)? If so, explain how this information helps determine the phylum.

4. *Physalia* is an example of a colonial organism. Explain what is meant by "colonial organism," and how *Physalia* fits this description.

5. What is one physical feature of the trematodes that helps distinguish them from the turbellarians?

6. What are two physical features of the cestodes that help distinguish them from the trematodes?

7. List some features that make the trematodes and cestodes adapted for a parasitic lifestyle.

8. Upon first glance, tapeworms may look like segmented worms. Why are tapeworms not classified as annelids? Contrast the proglottids of tapeworms with the segments found in segmented animals, such as the annelids.

9. What are three characteristic features of the body of a mollusk?

10. Where is the coelom located in mollusks?

11. Compare and contrast the oligochaetes and the polychaetes: List two similarities and two differences between these two classes.

12. List one physical characteristic of leeches that distinguishes them from polychaetes and oligochaetes.

13. What is a hydrostatic skeleton, and how does it function?

14. The arthropods and nematodes are ecdysozoans. What does this mean? (You may need to consult your text and lecture notes for further information.)

15. List two functions of the arthropod exoskeleton. What is a disadvantage of the arthropod exoskeleton?

16. As mentioned, nematodes have hydrostatic skeletons. What other phylum studied also possesses a hydrostatic skeleton?

17. State two ways in which an arachnid can be distinguished from an insect.

18. What are the four hallmark characteristics that all chordates possess during at least part of their lives?

19. How does the skeleton of a ray-finned fish differ in composition from that of a cartilaginous fish?

20. How does the amniotic egg make an organism adapted to terrestrial life?

21. What features make amphibians tied to a watery environment?

22. What features make reptiles completely adapted to a terrestrial environment?

23. What unique feature of mammals gives them their name? What is the function of this feature?

24. On a separate sheet, design a table that compares/contrasts the different animal phyla. Use the following as column headings in your table:

   A. Distinguishing characteristics: Visible features that help you distinguish this phylum from others
   B. Type of symmetry
   C. Number of embryonic tissues layers (diploblast/triploblast), where applicable
   D. Coelom present or absent? (NOTE: the terms acoelomate and coelomate only apply to triploblasts)
   E. Classes and/or subphyla seen in lab (where applicable)
   F. Representative organism(s)

FIGURE CREDITS

DIGESTION

All heterotrophs must digest organic molecules such as carbohydrates, fats, and proteins in order to make energy in the form of ATP. Digestion by heterotrophs either occurs intracellularly or extracellularly. **Intracellular digestion**, characteristic of most protists, occurs within the cytoplasm of cells. For example, the unicellular *Amoeba* engulfs food particles that then get broken down by enzymes within the organism. Other organisms such as the poriferans also carry out intracellular digestion.

Extracellular digestion occurs outside the cells of an organism. In this method of digestion, digestive enzymes secreted by cells digest food particles, and the by-products of the breakdown are then absorbed into cells. The enzymes are either secreted directly onto the food source, as seen in the fungi, or are secreted into the lumen of a digestive tract into which food has been ingested, as seen in chordates and many other animals.

Today's lab focuses on mammalian digestion. Figure 55 shows the main components of the human digestive tract; you will see many of these in the rat as well. During swallowing, food travels from the **oral cavity** into the **stomach** via the **esophagus**. The stomach, in addition to storing food, secretes gastric juice, containing a variety of substances including mucus, HCl, and the protease pepsin. Partially digested food moves from the stomach into the small intestine, where a majority of digestion and absorption occurs. Digestion occurs via integral membrane enzymes in the cells facing the lumen of the tract, as well as by enzymes secreted by the **pancreas** into the **duodenum**. The first portion of the **large intestine** is the **cecum**, a blind-ended sac. Herbivores such as the rat have a large cecum that contains bacteria capable of cellulose breakdown. In humans, the cecum functions mainly in water (and sometimes electrolyte) absorption. The **liver** has a many functions including the production of bile, which is important in fat emulsification. The **gall bladder** stores and concentrates bile (note that rats do not have a gall bladder).

Today you will investigate the activities of three of the many enzymes involved in mammalian digestion and will continue your dissection of the rat.

Pharynx

Salivary Glands
Parotid
Sublingual
Submandibular

Oral cavity

Uvula
Tongue

Esophagus

Liver
Gallbladder

Stomach

Pancreas
Pancreatic duct

Common
bile duct

Colon
Transverse colon
Ascending colon
Descending colon

Small Intestine
Duodenum
Jejunum
Ileum

Cecum
Appendix

Rectum
Anus

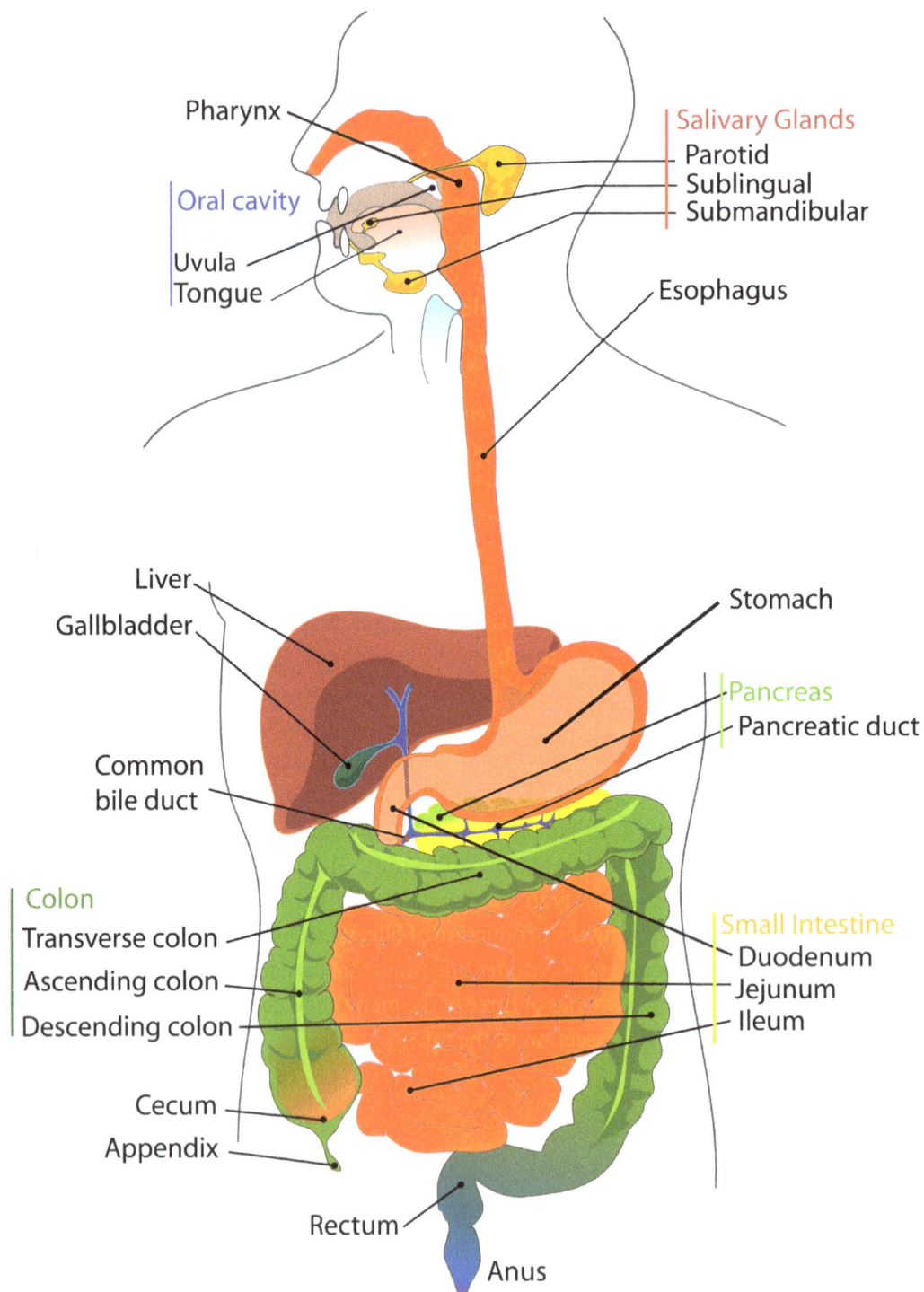

FIGURE 55. **The human digestive tract.**

ACTIVITIES

Exercise 1: Investigation of Lipid Digestion by Pancreatic Lipase

(experiment modified from Tharp and Woodman, 2008)

Pancreatic lipase is an enzyme secreted by the pancreas into the duodenum during digestion. It breaks down triglycerides into free fatty acids and glycerol and also into free fatty acids and monoglycerides. In today's experiment, you will investigate the digestion of fats by pancreatic lipase in the presence and absence of bile salts. Cream (or half and half) will be used as the source of fat; the cream will be pre-mixed with litmus powder, a chemical that is blue/purple in alkaline environments and pink in acidic environments. A specific color change (from blue to pink) will occur in the tubes in which lipase was working. Pancreatin, a mixture of pancreatic lipase, pancreatic amylase, and trypsin, will be used as the enzyme source. Note that although three enzymes are present, you are only investigating the activity of the lipase.

PROCEDURE

1. Prepare four test tubes as indicated in the table below. Once all components are in each tube, mix each tube by gently shaking.

TABLE 8. Reaction Tubes for Pancreatic Lipase Activity Experiments.

Tube #	Cream/Litmus Mixture	1% Pancreatin	Water	Bile Salts
1	3 mL	3 mL		
2	3 mL		3 mL	
3	3 mL	3 mL		a pinch*
4	3 mL		3 mL	a pinch

*Just a small amount of bile salts on the tip of the measuring spatula provided is enough.

2. Note and record the initial colors of the tubes in Table 9. Incubate tubes 1–4 at room temperature for 45 minutes. Note and record the colors of the tubes at 5 minutes, 10 minutes, 20 minutes, and 45 minutes in Table 9.
3. Please put your group's bottle of cream/litmus in the refrigerator when you have finished this experiment.
4. Answer the questions corresponding to this exercise, found at the end of this chapter.

TABLE 9. Color Changes Observed after Incubation of a Cream/Litmus Powder Mixture with Different Combinations of Pancreatin, Water, and Bile Salts.

Tube #	Initial color	Color after 5 minutes	Color after 10 minutes	Color after 20 minutes	Color after 45 minutes

Exercise 2: Colorimetric Test for Investigating Salivary Amylase Activity

Salivary amylase is an enzyme secreted by the salivary glands into the oral cavity. It breaks down polysaccharides, mainly starch, into oligosaccharides. It works optimally in a neutral to slightly acidic environment; most of its digestive activity occurs in the mouth, and when it reaches the stomach, salivary amylase is inactivated by the gastric juice, which has an average pH of 2.0.

PROCEDURE

Work in pairs. Each pair should obtain the following:
two graduated plastic transfer pipettes
two glass test tubes
squeeze bottle of distilled water
one plastic medicine cup
one dropper bottle iodine solution
one small bottle of 0.5% starch

1. Label the test tubes "1" and "2." Using a clean pipette, measure 1 mL 0.5% starch solution into each test tube.
2. Using a clean pipette, measure 2.5 mL water into test tube #1.
3. Generate some saliva: Think about your favorite food for a bit and then spit into a medicine cup—try to obtain 2.5 mL—if you cannot, dilute your saliva to the 2.5 mL marking on the cup.
4. Using a clean pipette, mix the saliva in the medicine cup and transfer all of the saliva mixture into test tube #2.
5. Add one drop of iodine solution into each tube. Observe the results and record your observations in Table 10.

TABLE 10. Digestion of Starch by Salivary Amylase: Experimental tube contents and results.

Tube #	Tube Contents	Color of tube contents
1		
2		

NOTE: Iodine + starch = blue color; if no starch is present, the solution will either be clear or will be yellowish-brown (the color of the iodine solution).

6. Answer the questions corresponding to this exercise, found at the end of this chapter.

Exercise 3: Investigation of Protein Digestion by Pepsin

(experiment modified from Tharp and Woodman, 2008)

Pepsin is an enzyme secreted into the lumen of the stomach by cells called chief cells found in the gastric glands. (It is actually secreted as an inactive form called pepsinogen, and is activated in the lumen to pepsin by HCl.) Pepsin is one of the many digestive enzymes that digest proteins; however, pepsin is the only protease found in the stomach. All other proteases are made by either the pancreas or small intestine.

Because pepsin is active in the stomach, in what type of environment do you think it works most effectively: alkaline, neutral, or acidic?

PROCEDURE

Because of the time required to observe results for these experiments, the reaction tubes were set up prior to lab and left to incubate in a 37°C water bath. The protein source used in this experiment was a small piece of boiled egg white (which is mainly albumin, a protein).

1. At your lab bench, you will find a set of four numbered test tubes in a test tube rack. Below is a list of the contents of each test tube at the beginning of the experiment. The specific tube number containing each set of contents is not given; you will need to assign each content list a tube number, based on your observations of the results.

Test tube # _____ :

boiled egg white
5 mL of 5% pepsin solution
5 mL water

Test tube # _____ :

boiled egg white
5 mL of 5% pepsin solution
5 mL 0.1M HCl

Test tube # _____ :

boiled egg white
5 mL 0.1M HCl
5 mL water

Test tube # _____ :

boiled egg white
5 mL of 5% pepsin solution
5 mL 0.1M NaOH

2. Put on gloves and observe what you see in each tube: Is the boiled egg white piece still visible or is it broken down/not visible? CAREFULLY determine the pH in each tube: Use forceps to obtain a piece of pH paper, and carefully tilt the tube until the paper touches the liquid in the tube. Do not drop the paper into the tube. Record your results in Table 11.
3. Given your observations and what you know about the optimal environment for pepsin activity, assign tube numbers to the content lists above.
4. Answer the questions corresponding to this exercise, found at the end of this chapter.

TABLE 11. Results of Incubation of Boiled Egg White with Different Mixtures of 5% Pepsin, Water, 0.1M HCl, and 0.1M NaOH

Tube #	Egg White Piece: Visible or Not Visible?	pH of Tube Contents
1		
2		
3		
4		

Exercise 4. Dissection of the Rat: Digestive and Respiratory Systems

Today you will continue your dissection of the rat by focusing on structures of the digestive and respiratory systems. Obviously, it is impossible to avoid looking at other structures, but note that structures not belonging to these systems will be focused on in subsequent labs. Refer to Appendix 2 and the description below for assistance with locating the structures listed in the table.

1. Put on gloves. Obtain your rat from last week, along with a dissecting tray, scissors, a scalpel, dissecting probe, and pins.
2. Using scissors, make a cut on the ventral side of the rat (through the body wall, not just the skin) that runs from the urethral opening to the base of the neck. Be sure to cut through the rib cage as you are cutting into the thoracic cavity. Next, make a cut just posterior to the diaphragm that is perpendicular to the first cut and that extends from the left to the right side.
3. Pin the body wall of the rat to the tray to expose the internal organs. The **diaphragm** is a skeletal muscle that divides the thoracic and abdominal cavities. The contraction and relaxation of the diaphragm allows for ventilation, the movement of air into and out of the lungs. The lobed **lungs** reside in the thoracic cavity, just anterior to the diaphragm. The heart is located near the center of the thoracic cavity, and will be discussed in a future lab. Continue cutting anteriorly toward the mouth and peel back the skin. Using forceps, pick away the skeletal muscle tissue in the neck to expose the **trachea**, an air passageway that is reinforced by cartilaginous rings. Just posterior to the diaphragm is the large multi-lobed **liver**, which has many functions including detoxification, storage of glycogen, processing nutrients, and the production of bile, which emulsifies fats in the small intestine. Gently lift the liver to reveal the **stomach**, a J-shaped organ functioning in stomach churning and storage, and in initial protein digestion via the protease pepsin. The **esophagus** transports food from the mouth into the stomach and is most easily viewed by finding the point at which it enters the stomach. Note that the diaphragm surrounds the esophagus. In the stomach, food is converted into a liquidy substance called chyme. Chyme passes from the stomach into the **duodenum**, the first portion of the **small intestine**. The small intestine, in particular the duodenum, is the site of most digestion and absorption in the digestive tract. Digestive enzymes of the small intestine include those that are bound to the cell membranes of intestinal cells and also enzymes that are secreted by the **pancreas** into the duodenum. The pancreas in the rat is a diffuse organ that is found in the mesentery between the stomach and duodenum. It is most easily viewed by lifting the stomach to expose the mesentery. In addition to producing digestive enzymes, the pancreas also makes and secretes the hormones insulin and glucagon that help regulate blood glucose. The small intestine leads into the **cecum**, a blind-ended sac that is the first part of the **large intestine** (colon). The cecum contains bacteria that ferment carbohydrates. In rats, the cecum is fairly large. Virtually no digestion occurs in the large intestine, but some absorption of water and vitamins occurs here.

Digestive and Respiratory Structures to Identify in the Rat

Structure	Function
trachea	
lungs	
diaphragm	
esophagus	
stomach	
small intestine	
duodenum	
large intestine (colon)	
cecum	
liver	
pancreas	

QUESTIONS

Exercise 1:

1. In which tube(s) did you notice a color change? Explain your results. Were these the results you expected to see?

2. Why would the color change to pink in tubes in which the reaction was occurring? (Hint: Think of the products resulting from the breakdown of triglycerides.)

Exercise 2:

3. Which tube was the control for this experiment?

4. Explain your observed results. Were they what you expected?

Exercise 3:

5. Thoroughly explain how you were able to assign tube numbers to each list of contents. Did any tubes act as a control?

FIGURE CREDIT

Figure 55: Source: http://commons.wikimedia.org/wiki/File:Digestive_system_diagram_en.svg. Copyright in the Public Domain.

BLOOD AND CIRCULATION

INTRODUCTION

While circulatory systems differ among animal groups, their function remains the same: to transport materials such as respiratory gases, nutrients, and wastes throughout the animal's body. In some animals, a fluid called hemolymph takes the place of blood. In today's lab, you will compare and contrast the blood and hearts of different animal groups, investigate the human ABO blood typing system, dissect the sheep heart, and continue your dissection of the rat.

ACTIVITIES

Exercise 1: Comparison of Amphibian, Reptilian, and Human Blood

Obtain prepared slides of amphibian blood, reptilian (both turtle and bird) Figures 56-58 blood, and human blood and observe under the compound microscope at your bench. Note that you may need to use the oil immersion lens to most clearly see the detail of the human blood smear. Draw what you see in each slide and label the erythrocytes and leukocytes (if present) in each.

Exercise 2: ABO and Rh Blood Typing

The four main human blood groups are A, B, AB, and O. These blood types are distinguished from one another according to the presence or absence of specific antigens on the erythrocyte surface. These antigens have been designated as A and B, and are two of the best-known erythrocyte antigens. As seen in Table 12, people with A antigens on their erythrocytes are said to have blood type A; people with B antigens on their erythrocytes have blood type B; people with both A and B antigens on their erythrocytes have blood type AB; and people with neither A nor B antigens on their erythrocytes have blood type O.

Antibodies against antigens A or B begin to build up in the blood plasma shortly after birth. Antibody types are named as follows:

Antibodies against the A antigen = anti-A antibodies
Antibodies against the B antigen = anti-B antibodies

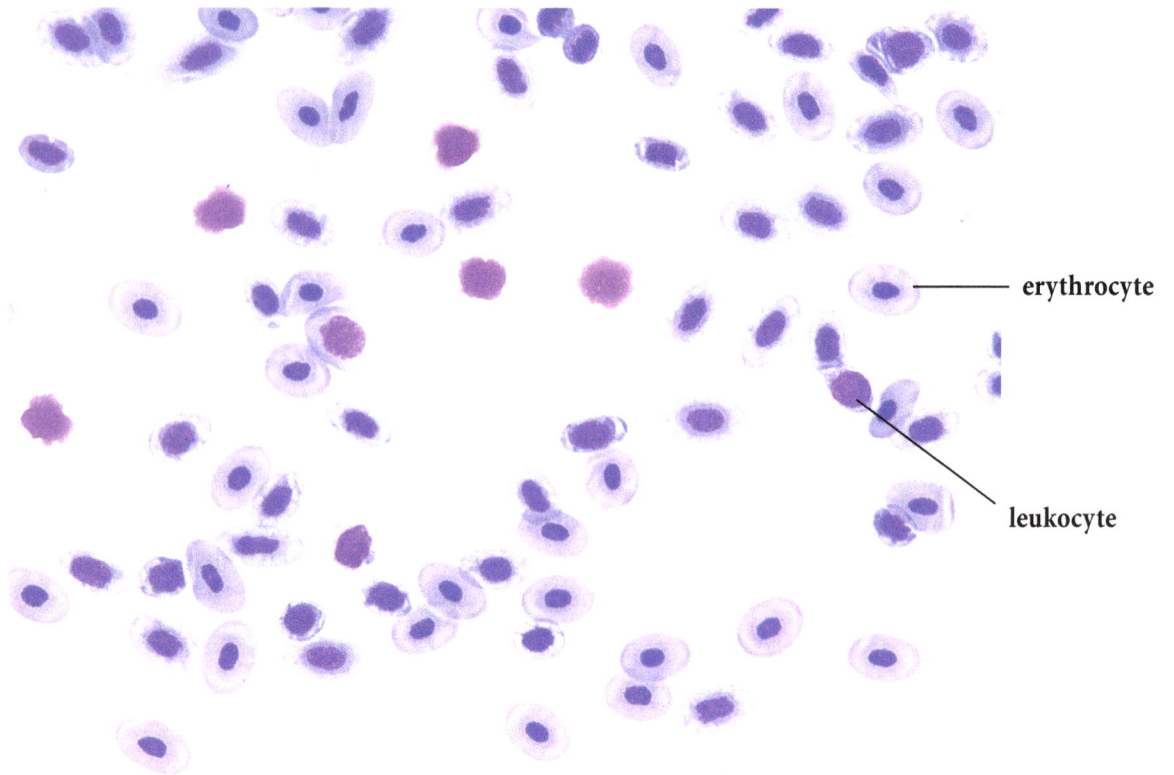

FIGURE 56. Frog blood smear, 471x.

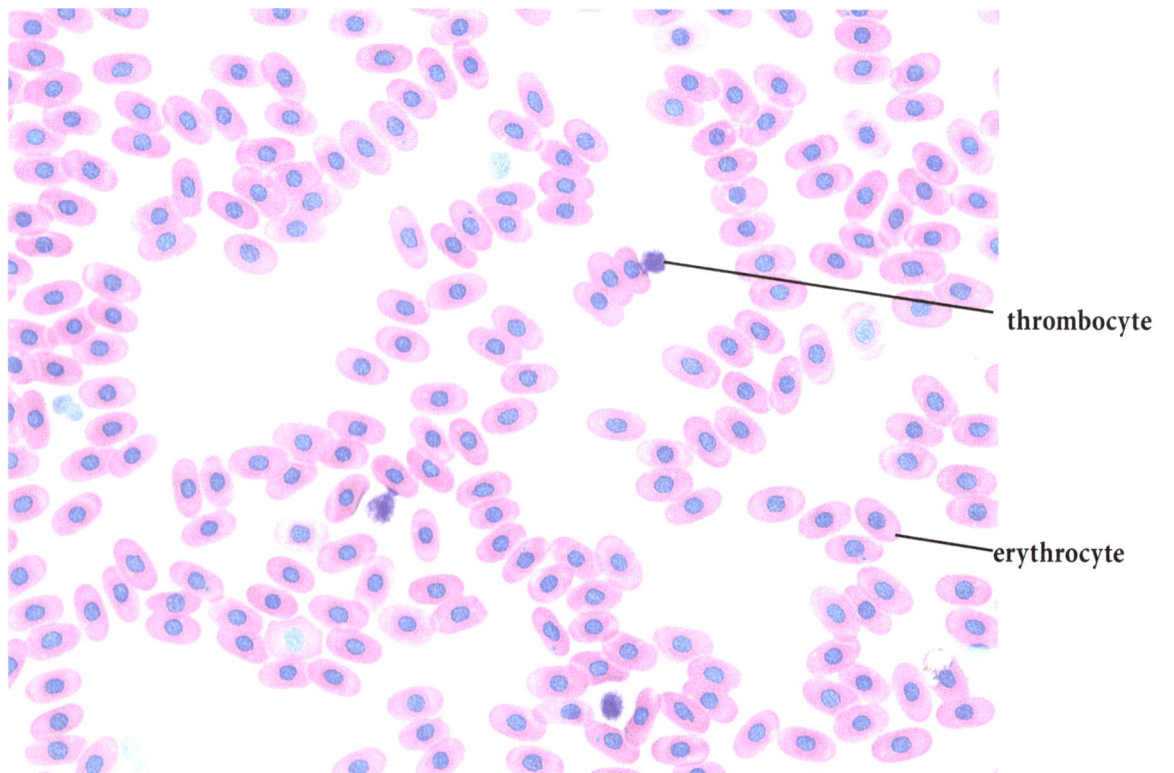

FIGURE 57. Turtle blood smear, 471x.

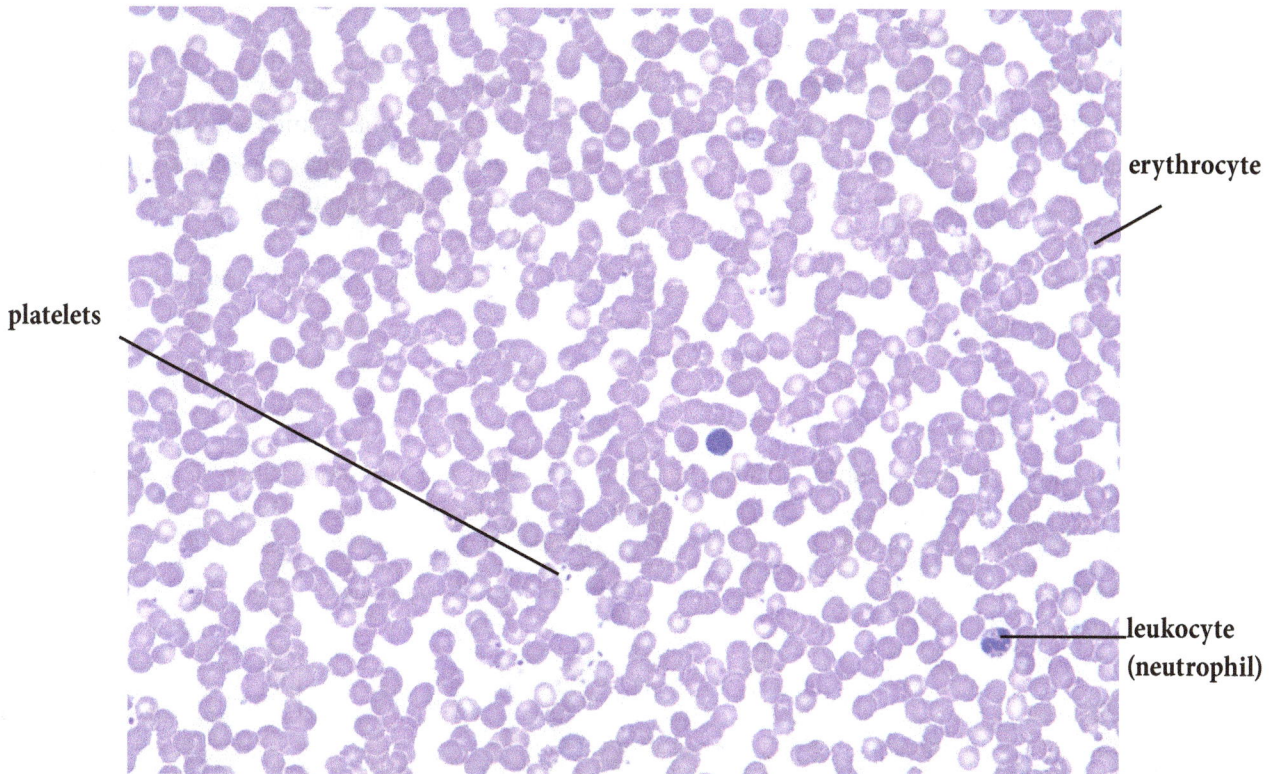

FIGURE 58. Human blood smear, 450x.

Humans produce antibodies against those antigens that are NOT on their erythrocytes. Always remember: Blood type is based on the antigens on the erythrocytes, not the antibodies in the plasma.

TABLE 12. **ABO Blood Groups.**

Blood Type	Antigens on Erythrocytes	Antibodies in Plasma	Can Give Blood to	Can Receive Blood from
A	A	Anti-B	A, AB	O, A
B	B	Anti-A	B, AB	O, B
AB	A and B	Neither Anti-A nor Anti-B	AB	O, A, B, AB
O	Neither A nor B	Both Anti-A and Anti-B	O, A, B, AB	O

As noted in Table 12, people can receive transfusions of only certain blood types, depending on the type of blood they have. If incompatible blood types are mixed, **agglutination** (clumping of red blood cells) occurs, followed by **hemolysis** (erythrocyte destruction). When performing transfusions, it is obviously optimal to match blood types of patient and donor. However, in certain situations where a smaller transfusion is necessary, it is possible to give a patient blood of a different type (Table 12). To make sure that agglutination does not occur, it is important to properly match the plasma of the patient and the erythrocytes of the donor. (Incoming plasma of the donor is sufficiently diluted in a small transfusion, so usually incoming antibodies in the donor plasma do not cause problems).

For more information on agglutination and hemolysis, refer to your textbook.

The Role of Agglutination in Determining Blood Type

Blood type can be determined easily by observing how a person's blood reacts with anti-A and anti-B antibodies. In this test, two small drops of blood are taken. To one sample, anti-A antibodies are added and to the other, anti-B antibodies are added. If agglutination occurs only in the suspension to which anti-A antibodies were added, the blood is type A. (Can you explain this?) If agglutination occurs only in the anti-B mixture, the blood is type B. Agglutination in both samples indicates that the blood type is AB. The absence of agglutination in any sample indicates that the blood type is O. These results are summarized in Table 13.

Rh Factor

The Rh factor is another surface antigen that can be found on erythrocytes. An individual who possesses this antigen on his or her erythrocytes is designated Rh+; an individual who lacks it is designated Rh-.

TABLE 13. Determination of Blood Type by Observing Blood Sample—Antibody Reaction.

Reaction of Blood with Anti-A Antibodies	Reaction of Blood with Anti-B Antibodies	Blood Type
Agglutination	No Agglutination	A
No Agglutination	Agglutination	B
Agglutination	Agglutination	AB
No Agglutination	No Agglutination	O

Anti-Rh antibodies are not normally present in the plasma, but anti-Rh antibodies can be produced in an Rh- individual upon sensitization to the Rh antigen. There are several ways sensitization can occur. For example, sensitization to the Rh antigen can occur if Rh+ blood is transfused into an Rh- recipient, or when an Rh- mother carries a fetus that is Rh+. Sometimes during childbirth, maternal and fetal blood mix as the placenta tears from the uterine lining. In this case, if the mother is Rh- and the fetus is Rh+, some of the fetal Rh antigens may enter the mother's circulation and sensitize her so that she begins to produce anti-Rh antibodies. Because it takes some time to build up the anti-Rh antibodies, the first Rh+ child carried by a previously unsensitized mother is usually unaffected. However, if an Rh- mother or a mother previously sensitized by a blood transfusion or a previous Rh+ pregnancy carries an Rh+ fetus, maternal anti-Rh antibodies may enter the fetal circulation, causing agglutination and hemolysis of fetal erythrocytes. This condition is called erythroblastosis fetalis.

The presence of the Rh factor can be tested for in the same manner described for testing for the presence of A and/or B antigens: anti-Rh antibodies are added to a drop of blood, and if agglutination occurs, the Rh factor is present (the person is Rh+).

PROCEDURE

Today you will investigate human blood types via carrying out a typing exercise on simulated blood or on your own blood. Please follow the directions of your instructor. If you are typing your own blood, follow the procedure below. If simulated blood is used, follow the protocol given in lab.

NOTE: When working with human blood, take appropriate precautions: Wear gloves when not pricking your finger, wipe any blood spills with bleach, and only work with your own blood. Dispose of blood-contaminated materials in the designated biohazard containers.

1. Obtain the following materials:

    Blood-typing tray
    sterile spring-loaded lancet
    alcohol pad
    Band-Aid
    anti-A, anti-B, and anti-D (Rh) antisera (these will be shared)
    three toothpicks

2. Get your supplies organized: Place the blood-typing tray in front of you on your lab bench, open the Band-Aid wrapper in case you need a Band-Aid, and note the location of the nearest red biohazard sharps container.

3. Wipe a fingertip with the alcohol pad and allow to air dry.

4. To obtain drops of blood, use the sterile lancet to prick the cleaned fingertip. You will not be able to see the lancet at any point, as it is located in a protected chamber and is spring-loaded. It will discharge and retract when you press the release button (this is located at the end opposite from the lancet). Press the lancet end to the cleaned area of your finger and firmly press the release button. **Immediately** put the lancet in the red biohazard sharps container on your lab bench.

5. Add one drop of blood to each well on the blood-typing tray.

6. Immediately add one drop of each antisera to its corresponding well: **Add anti-A to the "A" well, anti-B to the "B" well, and anti-D to the "D" or "Rh" well. DO NOT touch the dropper to your blood.** If it is too difficult to add these yourself (for example, if you need to put a Band-Aid on your finger), you can have a lab partner add these antisera to the wells.

7. Using a separate toothpick for each well, gently mix the blood and antisera in each well.

8. Observe to see whether agglutination occurs in the wells. It may take several minutes for agglutination to occur, especially in the Rh well.

9. Fill out the following table with the results of your blood typing experiment. You may use a lab partner's results if you chose to not prick your finger.

TABLE 14. **Results of Blood Typing Experiment.**

Well	Did agglutination occur?	What is your blood type?	What blood types other than your own can you receive?	To what blood types other than your own could you donate in small transfusions?
A				
B				
D (Rh)				

Exercise 3: Comparison of Animal Hearts

The hearts of different animals can differ in several ways, including the number of chambers present and whether or not the septa (walls) between the chambers are complete or incomplete. Furthermore, in animals with open circulatory systems, the heart may have perforations that allow the blood or hemolymph to leave the heart and bathe the tissues.

Observe the display comparing the hearts of a clam, crayfish, ray-finned fish, frog, bird, and human. For each animal, note whether the system is open or closed, how many chambers are present in the heart, and if more than one chamber is present, the number of atria and ventricles.

Clam (phylum_____, class_____):
Open or closed system?:

How many chambers are present in the heart?

If more than one chamber is present, how many atria and ventricles are present?:

Crayfish (phylum_____, subphylum or class_____):
Open or closed system?:

How many chambers are present in the heart?

If more than one chamber is present, how many atria and ventricles are present?:

Ray-finned fish (phylum_____, class_____):
Open or closed system?:

How many chambers are present in the heart?

If more than one chamber is present, how many atria and ventricles are present?:

Frog (phylum_____, class_____):
Open or closed system?:

How many chambers are present in the heart?

If more than one chamber is present, how many atria and ventricles are present?:

Bird (phylum_____, class_____):
Open or closed system?:

How many chambers are present in the heart?

If more than one chamber is present, how many atria and ventricles are present?:

Human (phylum_____, class_____):

Open or closed system?:

How many chambers are present in the heart?

If more than one chamber is present, how many atria and ventricles are present?:

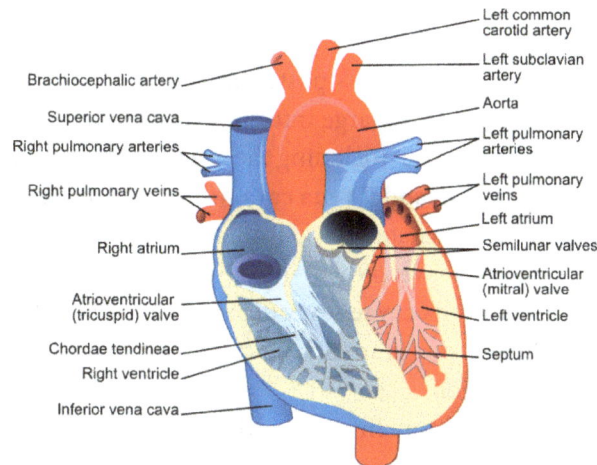

FIGURE 59. The human heart.

Exercise 4: Dissection of the Sheep Heart

1. Put on gloves. Observe the dissected sheep hearts on display and identify the following structures. Also refer to Figure 59 and to any heart models available in lab.

   right and left atria
   right and left ventricles
   aorta
   pulmonary trunk
   tricuspid valve
   bicuspid (mitral) valve
   semilunar valves

Exercise 5. Dissection of the Rat: Circulatory System

Continue your dissection of the rat, focusing on the circulatory system. Refer to Appendix 2 and the description below for assistance with locating the structures listed in the table.

1. Put on gloves. Obtain your rat from last week, along with a dissecting tray, scissors, a scalpel, dissecting probe, and pins.
2. Pin the body wall of the rat to the tray to expose the internal organs.

The four-chambered **heart** is located near the center of the thoracic cavity. The **aorta** is the main artery branching from the left ventricle. In double-injected specimens, the aorta will be pink. After exiting the left ventricle, the aorta curves, forming the aortic arch. There are three vessels that branch off the aortic arch: the brachiocephalic artery, the left common carotid artery, and the left subclavian artery. (The right common carotid artery splits off the brachiocephalic artery). The **common carotid arteries** supply oxygenated blood to the head, and can be seen running along either side of the trachea. Return your focus to the aortic arch. From here the aorta runs posteriorly along the dorsal side of the heart and into the abdominal cavity. Move the small and large intestines aside to see the descending aorta. Many arteries and arterioles branch off of the aorta, including the **renal arteries** delivering blood to the kidneys. The descending aorta runs next to the **caudal vena cava**, which will be blue in double-injected specimens. The caudal vena cava carries deoxygenated blood from the posterior parts of the body to the right atrium. Many veins drain into the caudal vena cava, including the **renal veins** that carry blood from the kidneys. The right and left **cranial vena cavae** carry deoxygenated blood from the head to the right atrium. (Note that in humans, there is one cranial vena cava.) The **external jugular veins** also carry deoxygenated blood from the head region.

The **pulmonary trunk** is the main vessel branching from the right ventricle. In double-injected specimens, the pulmonary trunk will be blue. The pulmonary trunk branches into the right and left **pulmonary arteries** that carry deoxygenated blood to the lungs.

The **spleen** is an organ that has functions in immunity and also in cleansing the blood. With respect to the latter, the spleen removes old and defective blood cells from circulation and also removes debris and foreign matter from the blood. In the fetus, the spleen is a site of erythrocyte production. The spleen is located in the left side of the abdominal cavity, lying just beneath the diaphragm and curling partially around the stomach.

Circulatory Structures to Identify in the Rat:

Structure	Function
heart	
aorta	
pulmonary trunk (branches into pulmonary arteries)	
carotid arteries	
external jugular vein	
cranial vena cava	
caudal vena cava	
renal veins and arteries	
spleen	

QUESTIONS

1. Which blood has anucleate erythrocytes?

2. What is a main function of erythrocytes?

3. What pigment makes human blood (and the blood of many other animals) red in color?

4. What is the general function of leukocytes in human blood?

5. What leukocyte type in humans is responsible for antibody production?

6. Arteries carry blood_____(away from, toward) the heart.

7. Veins carry blood_____(away from, toward) the heart.

FIGURE CREDIT

Figure 59: Copyright © ZooFari (CC BY-SA 3.0) at http://commons.wikimedia.org/wiki/File:Heart_diagram-en.svg.

REPRODUCTION

INTRODUCTION

Two types of reproduction are seen among animals: asexual reproduction and sexual reproduction. All animals carry out sexual reproduction, even those having asexual means of reproducing (for example, sponges and cnidarians). Sexual reproduction involves the production of haploid gametes via meiosis, and the subsequent union of gametes (fertilization) resulting in a diploid zygote. Through meiosis, sexual reproduction introduces genetic variation into the population. Asexual reproduction does not involve meiosis, and therefore does not introduce genetic variation. Today's lab focuses on sexual reproduction in vertebrates.

ACTIVITIES

Exercise 1. Gametogenesis

Gametogenesis in animals occurs via meiosis. In vertebrates, gametogenesis occurs within the gonads (the testes and ovaries). In today's lab, we will focus on mammalian gametogenesis.

1. Spermatogenesis

Spermatogenesis in the testes occurs within the walls of the seminiferous tubules, tightly packed tubules within the testes. Diploid stem cells called spermatogonia undergo meiosis and mature into haploid sperm. As the developing sperm progress in development, they migrate toward the lumen of the tubules so they can be released once mature (Figure 60).

A. Obtain a prepared slide of a cross-section of seminiferous tubules. Observe under the compound microscope and draw what you see. Label a seminiferous tubule and developing sperm cells.
B. Obtain a prepared slide of mammalian (rat or human) sperm and observe under the compound microscope. For best results, focus using oil immersion and the 100x objective.

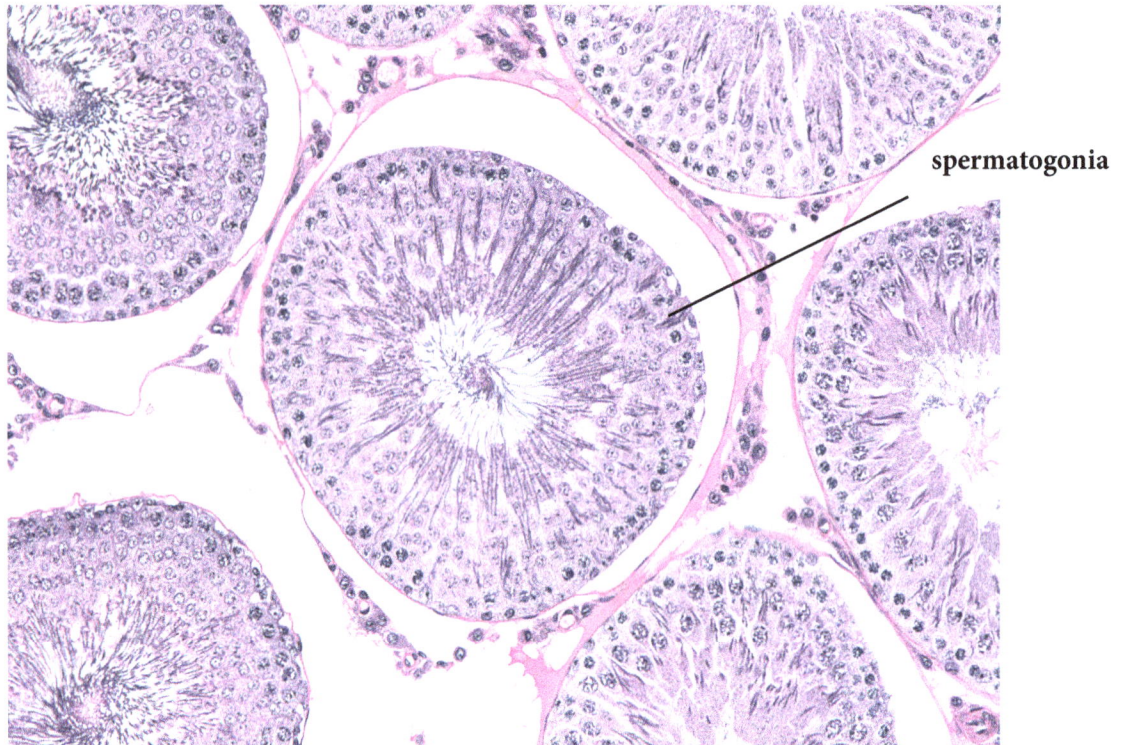

FIGURE 60. **Rat seminiferous tubules, 225x, c.s. Note the flagella of developing sperm in the lumen of the tubule.**

2. Oogenesis

Oogenesis in the ovaries begins with diploid stem cells called oogonia. Oogonia undergo meiosis and stages of maturation to become primary oocytes and eventually secondary oocytes. During the maturation process, each oocyte is surrounded by a ball of cells called follicular cells. Together, the oocyte and its surrounding follicular cells are called a follicle (Figure 61). The rupture of a mature follicle and release of the egg (secondary oocyte) into the Fallopian tube (uterine tube) is called ovulation. The ruptured follicle becomes a hormone-releasing structure called the corpus luteum. Consult your lecture and lab notes and textbook for more information on oogenesis and ovulation in mammals.

A. Observe the models of oogenesis on display and locate the follicles, oocytes, and corpus luteum.
B. Obtain a prepared slide of a section of a mammalian ovary showing follicles in different stages of maturation. Observe under the compound microscope and draw what you see. Label a follicle.

Exercise 2. Fertilization and Development in Vertebrates

Vertebrates exhibit either external or internal fertilization. External fertilization, the union of egg and sperm in the external environment, is found in the ray-finned fishes and amphibians. Fertilization occurring inside of the female reproductive tract is called internal fertilization, and is found in cartilaginous fishes, reptiles, and mammals.

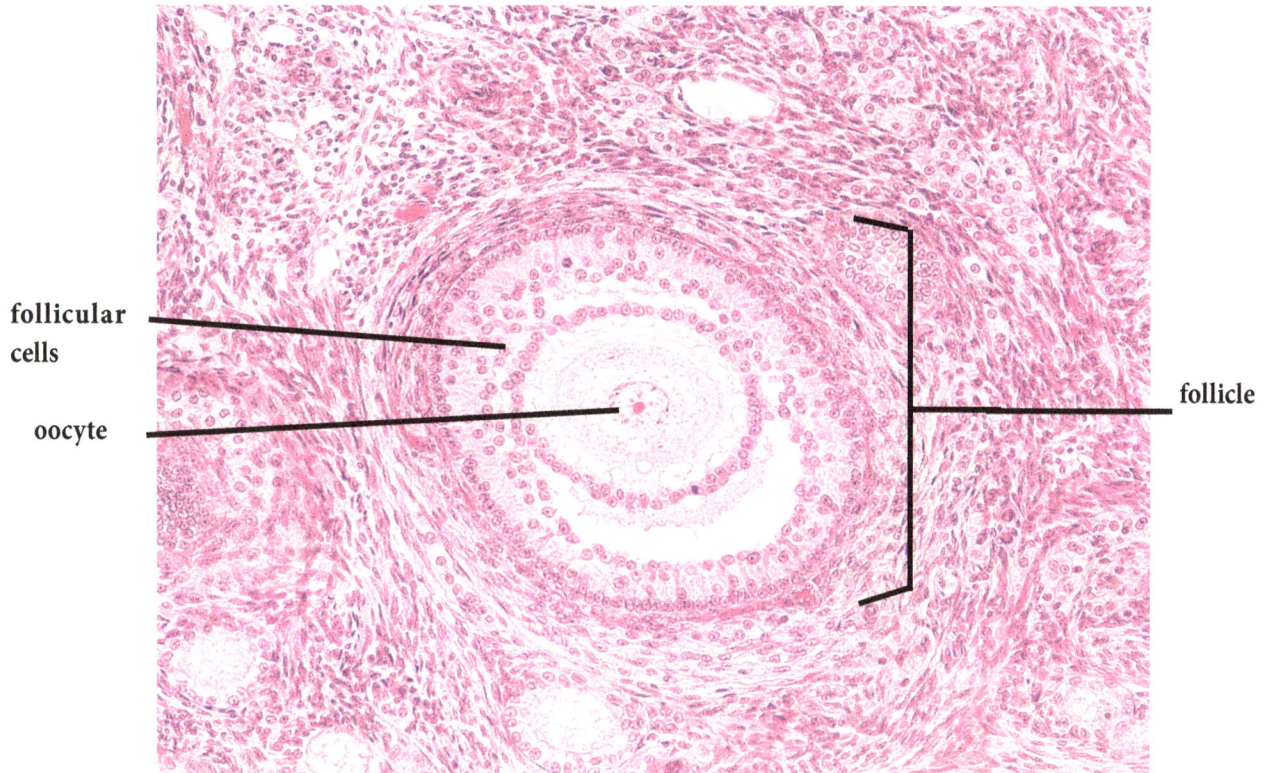

FIGURE 61. Developing follicle in a mammalian ovary, 225x.

Different patterns of reproduction are found among vertebrates. **Oviparous reproduction** is reproduction in which the embryo develops within an egg outside of the female's body. The embryo is nourished by yolk or a yolk sac within the egg, and eventually hatches from the egg. Oviparous reproduction is seen in the ray-finned fishes, amphibians, some cartilaginous fishes, and many reptiles (all birds and many others). The eggs produced by oviparous vertebrates differ in complexity and degree of protection. For example, the eggs of ray-finned fishes and amphibians are small and afford less protection than the egg of an oviparous shark. Oviparous reptiles and a few mammals (the platypus and echidna) produce amniotic eggs. As previously mentioned, the amniotic egg is an important adaptation that allows the embryo inside to safely develop in a terrestrial environment. The amniotic egg protects the embryo from desiccation, allows for sufficient gas exchange, and nourishes the embryo. Within an amniotic egg, the embryo is surrounded by a membrane called the **amnion**. The amnion and the fluid within it (amniotic fluid) provide protection and cushioning for the embryo. The yolk sac provides nutrients for the developing embryo, the chorion functions in gas exchange, and the allantois collects wastes from the embryo. The **extraembryonic membranes** described are also found in viviparous vertebrates within the uterus. Vertebrates possessing an amnion and other extraembryonic membranes during development are called **amniotes**.

In **ovoviviparous reproduction**, nourishment of the embryo still comes from yolk, but the embryo develops within an egg in the uterus of the mother until the embryo hatches and the mother gives live birth. This pattern of development is seen in some cartilaginous fishes and some reptiles. **Viviparous reproduction**, seen in some cartilaginous fishes, some reptiles, and most mammals, involves development of the embryo within the mother. The embryo obtains its nutrients via placental attachment in the uterus and the mother gives live birth.

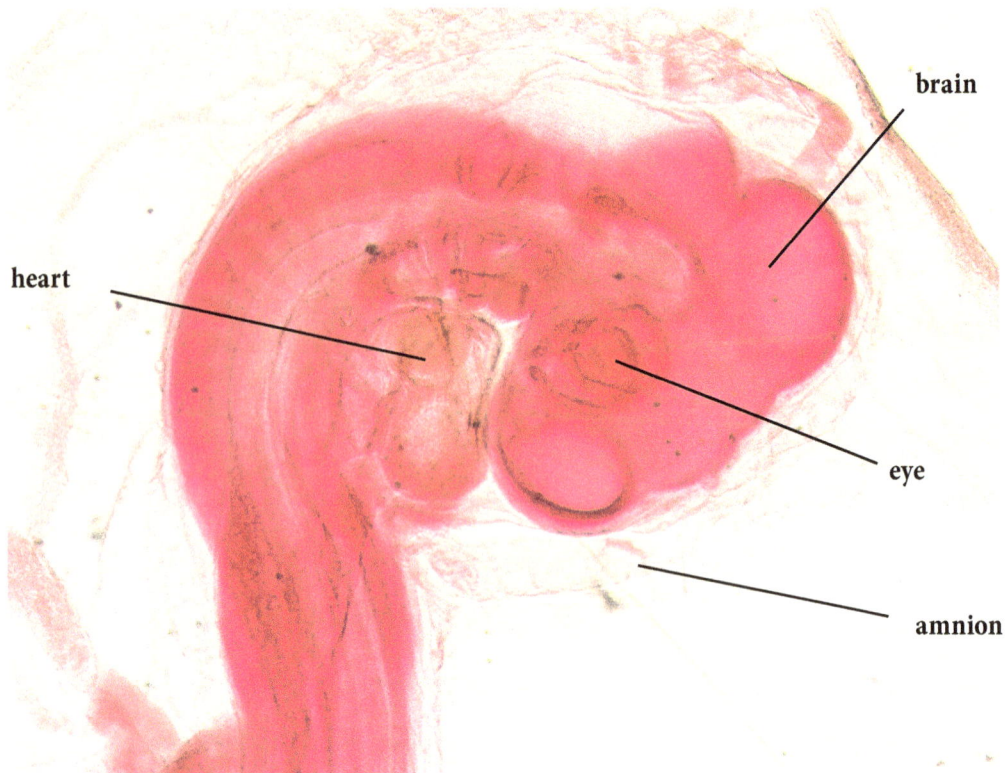

FIGURE 62. 72-hour chick, w.m., 22x.

1. Observe the display comparing fish, amphibian, and chicken eggs and note any differences and similarities between them.

2. Obtain a prepared slide of a whole mount of a 72-hour chick. Observe under the compound microscope at your bench. Identify the amnion, eye, brain, and heart. Refer to Figure 62.

3. Observe any preserved mammalian uteruses on display. Note any embryos present. Are you able to see any extraembryonic membranes?

4. In your textbook and in the lab, study diagrams of a human fetus developing in the uterus. Be able to identify the amnion, chorion/chorionic villi, placenta, and umbilical cord.

Exercise 3. The Human Reproductive System

1. Observe the model of the human male reproductive system. The **testes**, located within a sac called the scrotum, carry out spermatogenesis and make androgens, mainly testosterone. Surrounding each testis is a comma-shaped structure, the **epididymis**. The epididymis is a location for sperm storage and maturation. From each epididymis extends a **vas deferens** (ductus deferens), which is a pathway for sperm to travel from the epididymis to the urethra. The **prostate** and **seminal vesicles**, along with the **Cowper's glands**, produce seminal fluid.

2. Observe the model of the female reproductive system. The **vagina** leads to the **cervix**, the opening to the **uterus**. Branching from the uterus are the Fallopian (uterine) tubes, each of which leads to an ovary. The ovaries carry out oogenesis and make estrogens and progesterone. When an ovum (oocyte) is ovulated, it is swept into a Fallopian tube by fingerlike structures called fimbrae. The oocyte travels into the uterus and if it has been fertilized, it implants in the uterine lining.

3. Locate the following structures on the models. The urinary bladder, urethra, and ureters are included here because of their proximity to the reproductive structures.

Reproductive and Urinary Structures to Identify in the Human:

Structure	Function
testes	
epididymis	
vas deferens	
prostate	
seminal vesicles	
penis	
urethra (in males and females)	
urinary bladder (in males and females)	
ureters (in males and females)	
vagina	
uterus	
Fallopian tubes	
ovaries	
kidneys	

Exercise 4. Dissection of the Rat: Reproductive and Urinary Systems

Continue your dissection of the rat, focusing on the reproductive and urinary systems. Be sure to look at your rat and at a rat of the opposite gender to become familiar with both male and female reproductive structures. Refer to Appendix 2 and the descriptions given in exercise 3 for assistance with locating the structures in the table. Structures will look fairly similar to those in humans with one exception. Rats, along with many other non-primate mammals, possess a bifurcated (V-shaped) uterus that allows for implantation of multiple embryos at once.

1. Put on gloves. Obtain your rat from last week, along with a dissecting tray, scissors, a scalpel, dissecting probe, and pins.
2. Pin the body wall of the rat to the tray to expose the internal organs.
3. Locate the following structures and indicate their function, where applicable (note that their function in the rat is the same as for the human reproductive system).

Reproductive and Urinary Structures to Identify in the Rat:

Structure	Function
testes	
epididymis	
vas deferens	
prostate	
seminal vesicles	
penis	
urethra	
urinary bladder (in males and females)	
ureters (in males and females)	
uterus	
ovaries	
kidneys	

QUESTIONS

1. Contrast between external and internal fertilization. What is the difference between these types of fertilization?

2. List the four extraembryonic membranes found in amniotes and state a function of each.

3. Compare and contrast oviparity, ovoviviparity, and viviparity.

THE NERVOUS SYSTEM

INTRODUCTION

The nervous systems of organisms, although they vary in their degree of complexity, carry out the same functions: they sense stimuli and coordinate appropriate responses to these stimuli. Nervous tissue is excitable tissue, and generates electrical signals in response to stimulation. This ability to conduct electrical signals or messages is what enables nervous tissue to carry out its sensory and response functions.

Nervous tissue is found in all animals except sponges. Two main types of nervous systems seen in animals are nerve nets and central nervous systems. A nerve net consists of a network of diffuse, interconnected nerve cells. Animals with nerve nets are in general radially symmetrical and have no brains or cerebral ganglia. Cnidarians are examples of animals with nerve nets. Bilaterally symmetrical animals have central nervous systems, in which neurons are clustered into cerebral ganglia or a brain at the anterior end. The evolution of this concentration of nervous tissue and eventually of sensory organs at the anterior end of organisms is called cephalization. Cephalization is advantageous to bilaterally symmetrical animals as it allows them to sense their environment as they enter it.

Today's lab focuses on the mammalian nervous system, composed of the central nervous system (CNS), which is the brain and spinal cord, and the peripheral nervous system (PNS), consisting of neurons carrying information to and from the CNS. Neurons carrying electrical messages toward the CNS are called sensory (or afferent) neurons, and those carrying electrical messages away from the CNS are called motor (or efferent) neurons. Sensory neurons, as their name suggests, sense many types of stimuli such as light, pressure, and pain, and transmit messages to the CNS, which integrates this information and coordinates an appropriate response. Motor neurons carry electrical messages from the CNS to effectors, or targets, throughout the body, which carry out a response to the initial stimulus. For example, some motor neurons carry messages to skeletal muscles, which stimulate them to contract and cause movement.

The special senses are vision, hearing, taste, and olfaction (smell). All of these involve organs containing specialized sensory receptors (nervous tissue) that are sensitive to light (vision), sound waves (hearing), and chemicals in solution (taste and olfaction). When stimulated, these receptors cause nerve impulses to be sent to the CNS, which then processes the signals received. Some of today's activities focus on two sensory organs: the eye and the ear.

ACTIVITIES

Exercise 1. Investigation of Neuron Structure

Neurons are nerve cells that conduct electrical signals called action potentials, or nerve impulses. Figure 63 shows a typical neuron and its parts (note that the shape seen here is one of several shapes of neurons that exist). The cell body is where the nucleus and other organelles are located. In addition to carrying out typical cell functions, the cell body synthesizes neurotransmitters, the chemical messengers of the nervous system. Neurotransmitters are stored in and secreted from the axon terminals. Dendrites are the receiving end of a neuron; when a stimulus reaches a dendrite, a current is generated on the cell membrane and travels along the membrane toward the cell body and axon. When the current reaches the axon, a nerve impulse (action potential) is generated and propagated along the length of the axon toward the axon terminals. When the impulse reaches the terminals, the neurotransmitter molecules are secreted via exocytosis. The axons of some neurons in mammals are covered by a myelin sheath, a lipid and protein wrapping that electrically insulates the axon and speeds up conduction of the impulse.

1. Examine the model of the neuron on display and identify the cell body, dendrites, axon, myelin sheath, and axon terminal(s). In the space below, indicate the functions of these structures. Consult your lecture and lab notes and your textbook.

Cell body: Myelin sheath:

Dendrites: Axon terminals:

Axon:

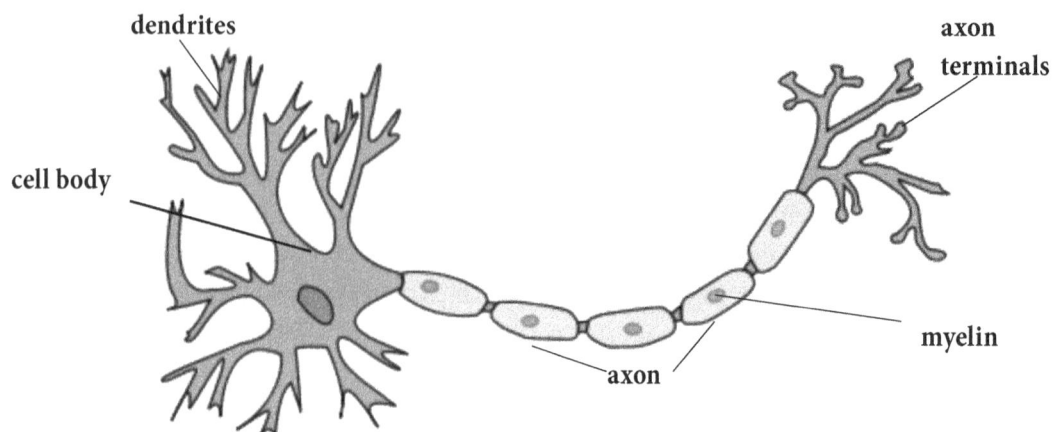

FIGURE 63. Diagram of a neuron.

2. Obtain a prepared slide of a smear taken from an ox spinal cord. Observe under the compound microscope and draw what you see. Label the cell body, nucleus, axon, and dendrites (it can be difficult to distinguish between axons and dendrites in these preparations).

Exercise 2. Dissection of the Mammalian Brain

The sheep brain will be observed as a representative mammalian brain.

1. Put on gloves. Obtain a dissecting tray, a whole brain, and a brain that has been cut along the sagittal plane.
2. Refer to Appendix 2 and the description that follows as you examine the brain to help you identify the structures listed in the chart.

The **cerebrum** is the largest portion of the brain and is generally responsible for conscious thought and memory. The **corpus callosum** can be seen in longitudinal section; it connects the right and left cerebral hemispheres and allows for communication between them. Posterior to the cerebrum is the **cerebellum**, which coordinates movement. The **pons** and **medulla oblongata**, along with the mesencephalon, comprise the brainstem. The brainstem relays information between the brain and spinal cord, controls ventilation, and regulates heart rate via control of the autonomic nervous system, among other functions. On the ventral side of the brain, the **optic chiasm** and **olfactory bulbs** are visible. The optic chiasm is the place where the optic nerves cross, and the olfactory bulbs are involved in olfaction.

3. When finished with your examination of the sheep brain, return your specimens to the area indicated by your lab instructor.

Structures to identify in the sheep brain:

Structure	Function
cerebrum	
cerebellum	
pons	
medulla oblongata	
optic chiasm	
corpus callosum	
olfactory bulbs	

Exercise 3. Investigation of a Somatic Reflex

Reflexes are involuntary responses to stimuli that are mediated by the spinal cord. Somatic reflexes are those that involve skeletal muscles. For example, the withdrawal of a hand from a hot stove or the retraction of a foot after stepping on a tack are somatic reflexes, since they involve neurons stimulating skeletal muscles to contract.

In today's lab, you will study a reflex commonly tested by physicians, the patellar (or "knee-jerk") reflex. The patellar reflex is an example of a stretch reflex. Tapping the patellar tendon stretches the quadriceps muscle in the thigh and stretches it. This stretch sends signals to the spinal cord, which in turns sends messages along neurons that fire onto the quadriceps and cause it to contract. This contraction raises the lower portion of the leg. Consult the diagram in the lab for more information on the patellar reflex.

1. Work in pairs. Obtain a reflex hammer and decide which person will be the "patient."
2. The patient should sit on the lab bench with his or her legs dangling off the edge of the table.
3. The tester should find the location of the patellar tendon (ask your lab instructor for help) and then tap the tendon with the reflex hammer. Observe the response. If it is difficult to elicit a response, distract the patient by asking him or her to count backwards, and then tap the tendon again.

VISION

When light enters the eye, it is focused by the lens on the retina, which contains specialized neurons and light-detecting (photoreceptor) cells called rods and cones. These allow humans to see black and white (rods) and color (cones). The focusing of light on the retina initiates events first in the photoreceptors and then a series of nerve impulses is sent to the brain via the optic nerve (Freeman, 2011). The processing of these impulses by the brain enables us to see images. The only place on the retina that lacks rods and cones is the point at which the optic nerve exits the eye. When light is focused on this area of the retina, there are no rods or cones to sense it, and the image is not seen. This area is called the optic disc, or blind spot.

FIGURE 64. Blind spot test.

Exercise 4. Determination of the Blind Spot

1. Hold Figure 64 at arm's length, with the star image located directly in front of your right eye. Cover your left eye and slowly move the image toward you, focusing your right eye on the star figure. When the circle disappears from your peripheral view, you have located your blind spot. Repeat this procedure for your left eye.

Exercise 5. Dissection of the Mammalian Eye

The sheep eye will be dissected as a representative mammalian eye.

1. Put on gloves. Obtain a dissecting tray, scissors, forceps, dissecting probe, and preserved sheep eye.
2. Before making any cuts, examine the diagram of the eye in Appendix 2 to become familiar with some main structures. The **sclera** is the tough protective outer tissue of the eye. What we know as the "white" of the eye is the sclera. The **cornea** is a transparent tissue that junctions with the sclera and allows light to enter the pupil (Marieb and Hoehn, 2010). The cut **optic nerve** extends from the posterior portion of the eye. At the anterior end, locate the **pupil** and **iris**. The iris is the colored part of the eye, which can constrict and dilate to regulate the amount of light going through the pupil and passing through the lens. In preserved specimens, it may be difficult to distinguish between the pupil and iris.
3. Place the eye on your tray and hold it firmly in place with one hand. With the other hand, carefully use the pointed tip of the scissors or a pointed dissecting probe to make a hole in the sclera. Once you have created a hole, cut into the sclera, ***making sure to cut away from you***. Be extremely careful while cutting the sclera, as the scissors may slip. Cut a circle around the outermost part of the iris (just to the outside of it) and remove the anterior portion of the sclera.
4. Locate the **vitreous humor**, the gelatinous substance that most likely spilled out of the posterior chamber of the eye as you cut the sclera. The vitreous humor transmits light, holds the retina in place, supports the lens, and provides pressure.
5. Identify the **lens** and **retina**. The retina is a delicate tissue covering the choroid. Many animals, including sheep, cats, and dogs, have a pearly reflective tissue called the tapetum lucidum just beneath the retina. This structure reflects light back through the retina and aids night vision.
6. Before discarding your dissected specimen, be sure you are able to identify the structures in the chart below.

Structures to identify in the sheep eye:

Structure	Function
sclera	
optic nerve	
iris	
pupil	
vitreous humor	
lens	
retina	

HEARING

The physiology of hearing involves sound waves entering the ear canal and causing vibrations of the tympanic membrane (eardrum). Upon vibration of the tympanic membrane, the incus, malleus, and stapes vibrate, thus causing vibration of the oval window. This causes waves in the cochlear fluid. Stereocilia on hair cells in the cochlea are bent by these waves, and impulses are sent along sensory neurons leading into the auditory nerve and to the brain.

Exercise 6. The Human Ear and Hearing

1. Examine Figure 65 and the model of the human ear on display. Identify the **auditory canal**, **tympanic membrane**, **incus**, **malleus**, **stapes**, **oval (elliptical) window**, and **cochlea**.
2. Obtain a tuning fork. Strike the fork firmly on your palm to initiate vibration and hold the fork to your ear. The sound waves of the fork are causing vibration of your tympanic membrane. The incus, malleus, and stapes vibrate, thus causing vibration of the oval window. This causes waves in the cochlear fluid. Stereocilia on hair cells in the cochlea are bent by these waves, and impulses are sent along sensory neurons leading into the auditory nerve and to the brain.
3. Strike the fork again, and this time touch the handle of the fork to the top of your skull. Can you hear the sound?

Sound waves conducted through air are normally heard better than those conducted through bone. However, in those with **conduction deafness**, sound waves conducted through bone are heard better. Conduction deafness is deafness resulting from an inability of sound waves to vibrate the tympanic membrane. Causes of conduction deafness include blockage of the ear canal (for example, by wax or your finger), perforation of the tympanic membrane, middle ear infection, and overgrowth of the ossicles. **Sensorineural deafness** results from damage to hair cells or neurons involved in hearing.

4. Simulate conduction deafness as follows. Strike the tuning fork, plug one ear, and touch the handle of the fork to the top of your skull. In which ear is the sound louder?

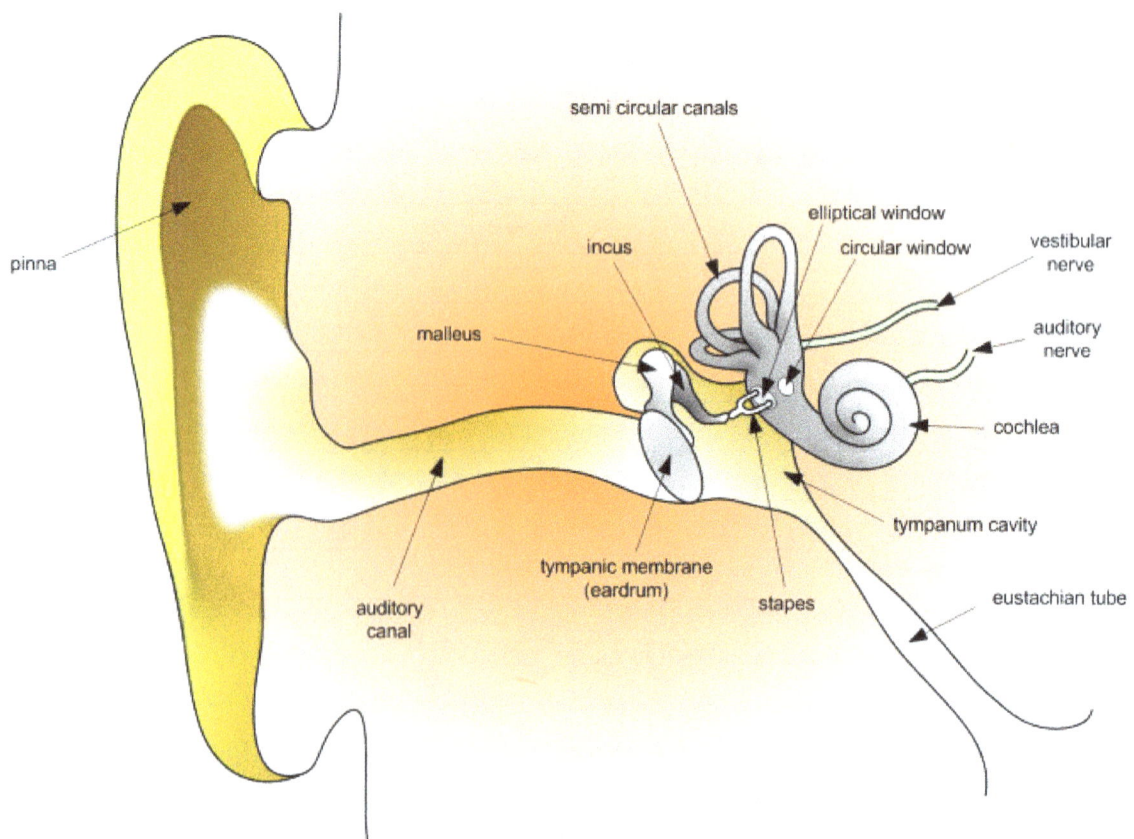

FIGURE 65. The human ear.

QUESTIONS

1.   Briefly describe a nerve net.

2.   What does the term "cephalization" mean? What is an advantage of having cephalization?

3   A.   When the patellar tendon is tapped, electrical messages are sent to the spinal cord. What type of neuron carries messages toward the CNS?

    B.   What type of neuron carries electrical messages away from the spinal cord to the quadriceps muscle?

4.   When a light focuses on the blind spot of the retina, the image cannot be seen. Why? What is another name for this area of the retina?

5.   Distinguish between conduction deafness and sensorineural deafness.

FIGURE CREDITS

REFERENCES

Banks, J. (1999). Gametophyte Development in Ferns. *Annual Review of Plant Physiology and Plant Molecular Biology* 50, 163–86.

Miller, S. M. (2010). *Volvox, Chlamydomonas,* and the Evolution of Multicellularity. *Nature Education* 3(9), 65.

Roberts, L., and J. Janovy. (2009). *Gerald D. Schmidt & Larry S. Roberts' Foundations of Parasitology.* New York: McGraw-Hill.

Sherman, I. W., and V. G. Sherman. (1976). *The Invertebrates: Function and Form: A Laboratory Guide. (2nd ed.).* New York: Macmillan Publishing Co., Inc..

Tharp, G. D., and D. A. Woodman. (2008). Digestion. In *Experiments in Physiology (9th ed.).* (pp. 103–4). San Francisco: Pearson Benjamin Cummings.

APPENDIX 1. MICROSCOPE USE REVIEW

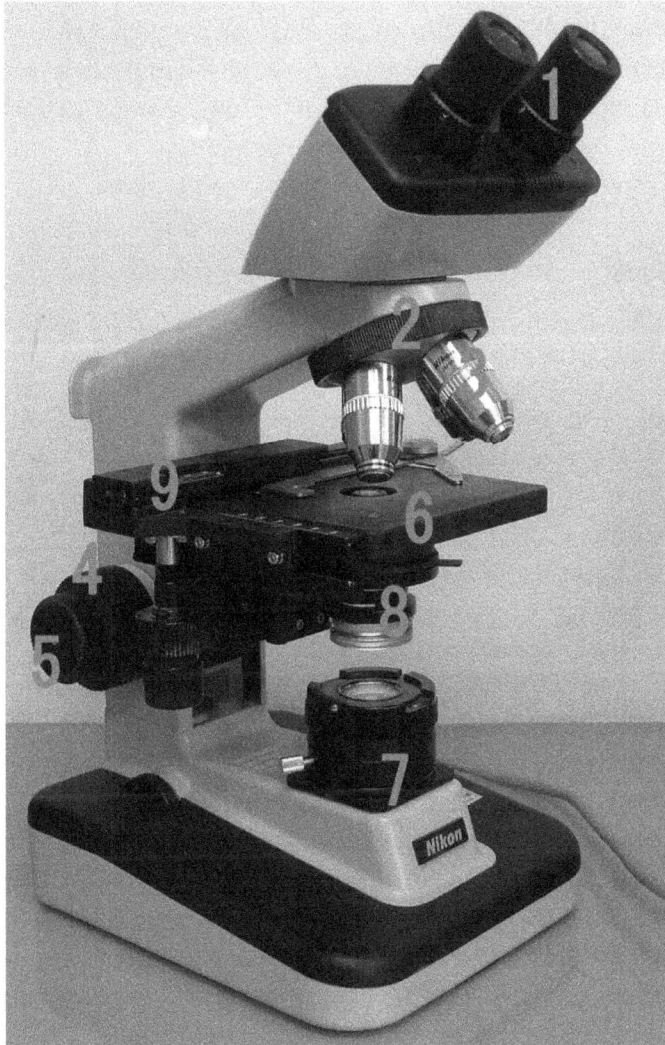

FIGURE A1.1 **Components of a compound light microscope:**

1. Ocular lens
2. Revolving nosepiece
3. Objective lens
4. Coarse focus knob
5. Fine focus knob
6. Stage
7. Lamp

8. Condenser lens with iris diaphragm
9. Stage adjustment knobs

General reminders for proper microscope care and use:

1. Always carry the microscope with two hands: one holding the arm and the other supporting the base.
2. Only lens paper should be used to clean lenses.
3. Immersion oil should only be used with the 100x objective lens. The other objectives are not sealed, and use of oil with them will result in oil seepage into the lens barrel.
4. The coarse focus knob should only be used with the lower power (4x, 10x) objective lenses. The fine focus knob may be used with all objective lenses.
5. The microscope should be stored with the 4x objective in position.

Proper focusing technique:

(NOTE: This is a general description to simply remind you how to focus a specimen. For a description of Kohler illumination setup to be done prior to microscope work, consult your lab instructor.)

1. Place the slide to be viewed on the stage and secure it in position. For best results, move the condenser close to the stage using the condenser adjustment knob.
2. Adjust the interpupillary distance between the oculars so that you may view the specimen comfortably with both eyes.
3. Always begin focusing using the lowest power objective, often the 4x. Use the coarse focus knob to bring the specimen into focus. Use the fine focus knob to fine tune the focus, if necessary. You may also wish to adjust the condenser lens and/or the iris diaphragm to control the amount of light hitting the specimen.
4. If you wish to increase magnification, gently rotate the revolving nosepiece so that the 10x is now in position. DO NOT move the stage prior to doing this. Look at the specimen and use the fine focus knob to bring the specimen into focus. Note that is it fine to use the coarse focus with the 10x, but you should not need to if the specimen has already been focused at 4x.
5. If you wish to increase magnification further, gently rotate the nosepiece so that the 40x objective is in position. The 40x objective will be very close to the slide, but will not touch it as long as the specimen has been properly focused at lower powers.
6. If you wish to view the specimen using the 100x objective, you must use immersion oil. Rotate the nosepiece and move the 100x objective almost in place, but not fully. Place a drop of immersion oil directly on the coverslip and then gently rotate the 100x lens into position. It should touch the oil. Use the fine focus knob to bring the specimen into focus. When finished, clean the 100x lens and the slide with lens paper.

FIGURE CREDIT

APPENDIX 2. DISSECTION GUIDES

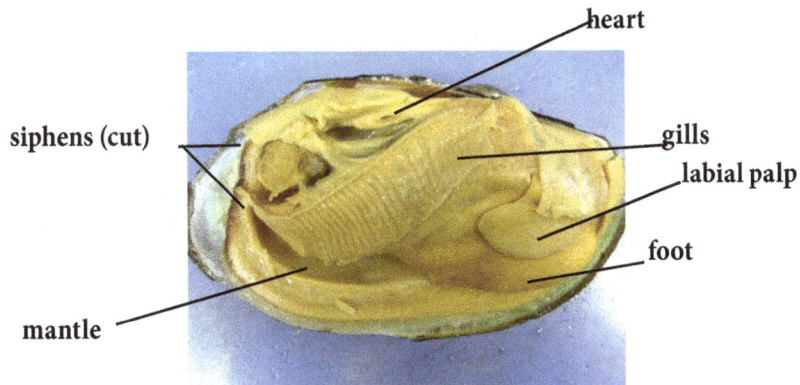

FIGURE A2.1. Freshwater clam. The right valve has been removed; the anterior end of the clam is on the right.

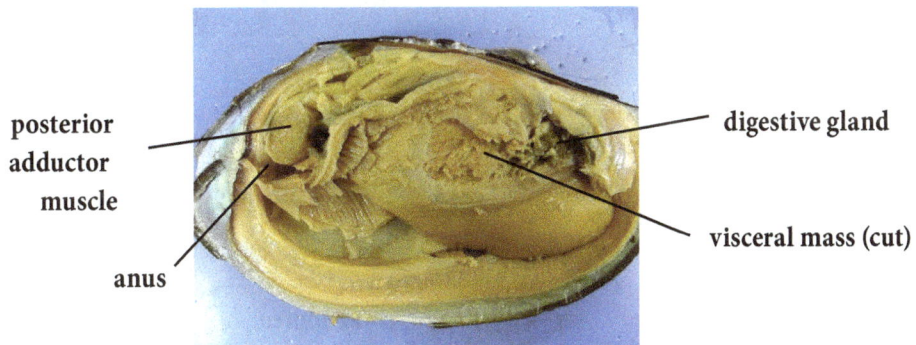

FIGURE A2.2. Freshwater clam. The gills and labial palps have been removed and the visceral mass has been sliced longitudinally.

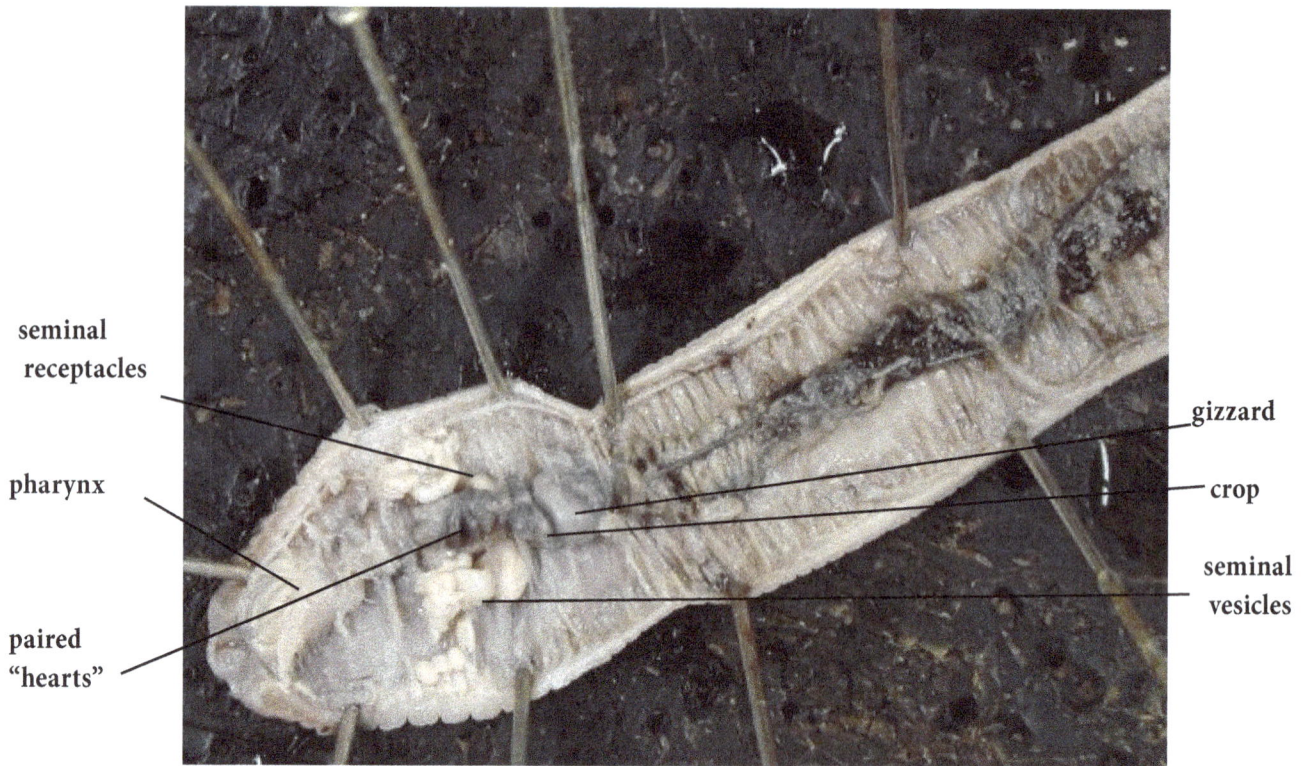

seminal receptacles

pharynx

paired "hearts"

gizzard

crop

seminal vesicles

FIGURE A2.3 **Earthworm. The animal has been cut along the dorsal side.**

cheliped

carapace covering the cephalothorax

tail

walking legs

abdomen

FIGURE A2.4. **Crayfish external anatomy.**

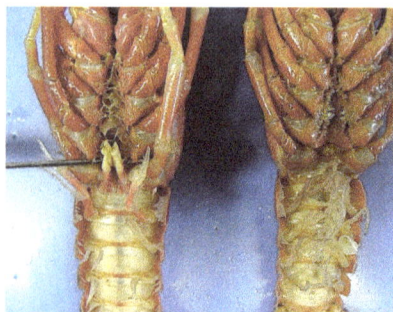

Copulatory swimmerets of the male are indicated by the probe.

FIGURE A2.5. **Crayfish male (left) and female (right), ventral view.**

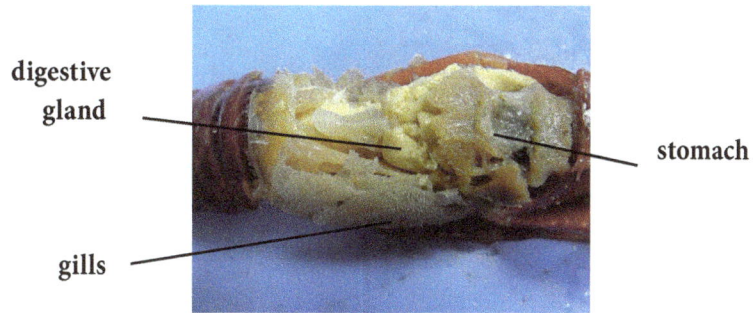

FIGURE A2.6. Crayfish internal anatomy, dorsolateral view.

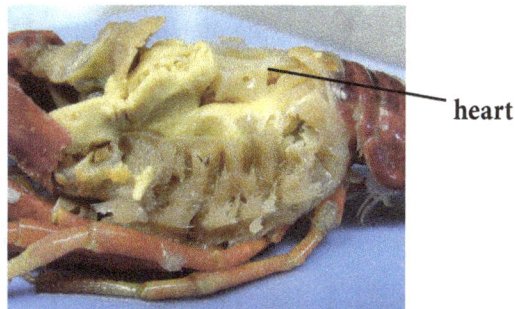

FIGURE A2.7. Crayfish internal anatomy, lateral view.

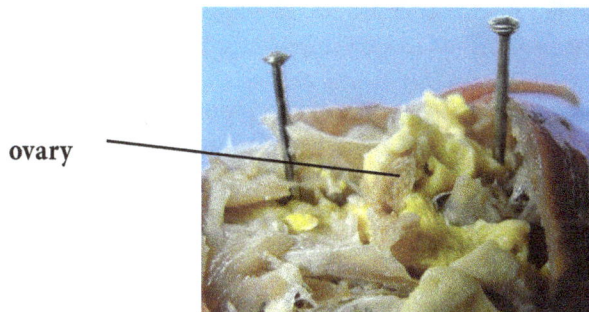

FIGURE A2.8. Crayfish internal anatomy, gravid female. Note the ovary filled with eggs. The heart and a portion of the digestive gland have been removed.

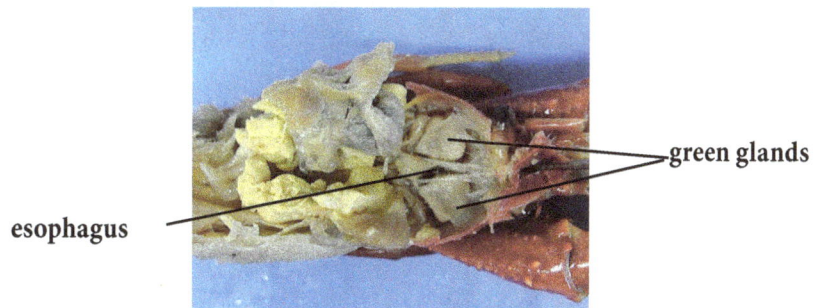

FIGURE A2.9. Crayfish internal anterior anatomy, dorsal view.

FIGURE A2.10. Grasshopper internal anatomy (female), dorsolateral view.

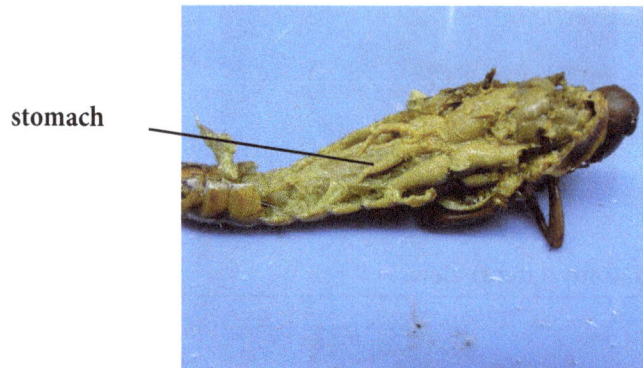

FIGURE A2.11. Grasshopper internal anatomy (female, ovaries removed), dorsolateral view.

FIGURE A2.12. Grasshopper internal anatomy, showing gastric caecae and Malpighian tubules.

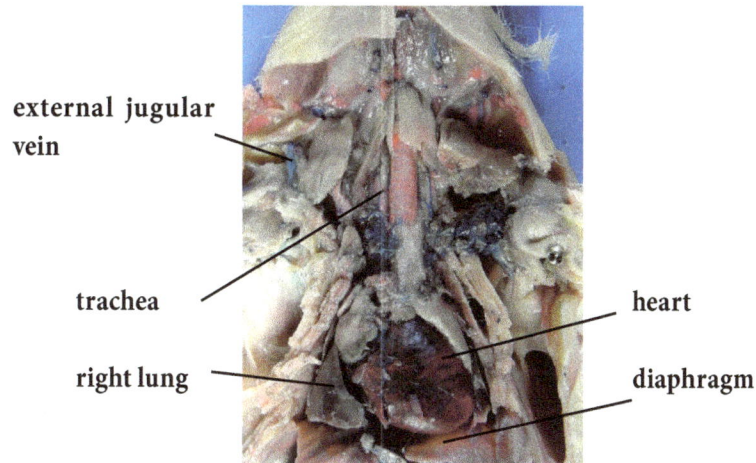

FIGURE A2.13. Rat internal anatomy, thoracic cavity.

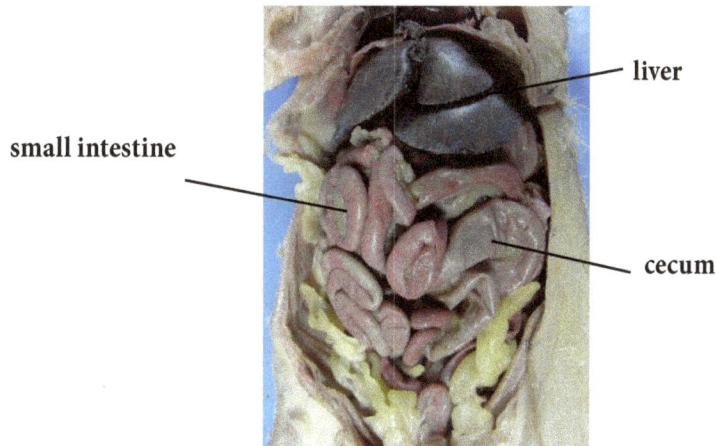

FIGURE A2.14. Rat internal anatomy, abdominal cavity.

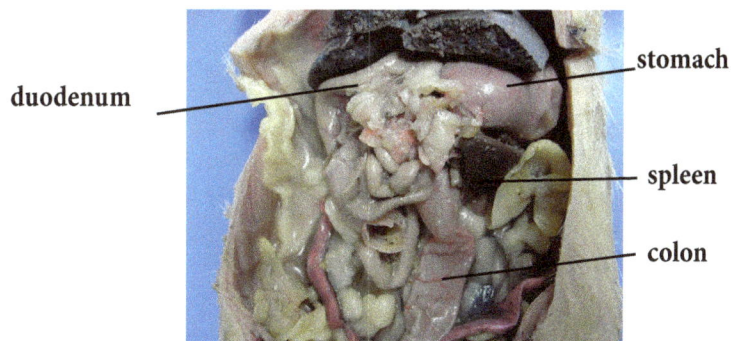

FIGURE A2.15. Rat internal anatomy, small intestine (jejunum and ileum) and most of large intestine removed (descending colon remains).

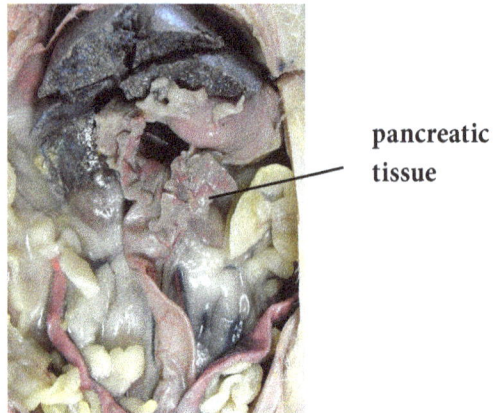

FIGURE A2.16. Rat internal anatomy, showing the location of the pancreas. Pancreatic tissue is found near the pyloric sphincter, the junction between the stomach and duodenum.

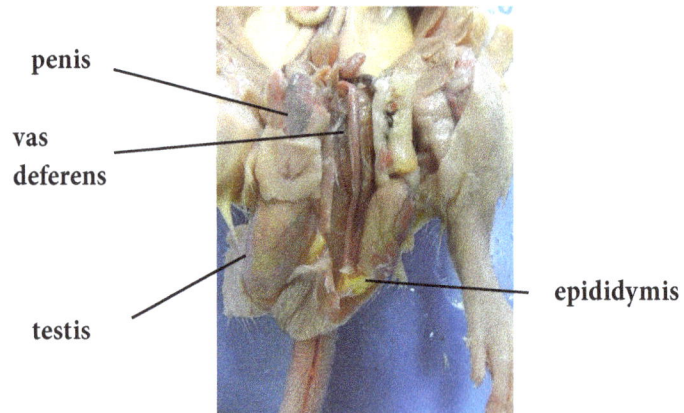

FIGURE A2.17. Internal reproductive anatomy of the male rat.

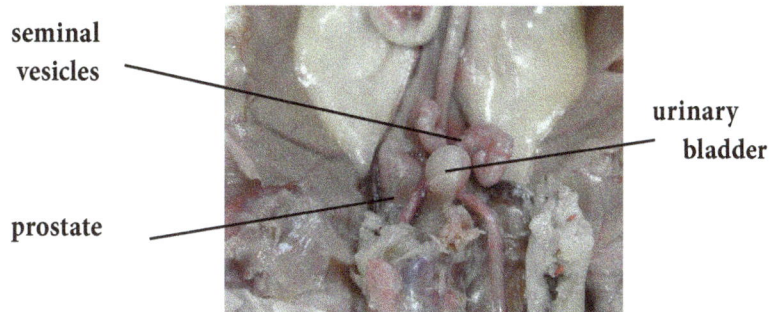

FIGURE A2.18. Internal reproductive anatomy of the male rat, showing the locations of the seminal vesicles and prostate gland in relation to the bladder.

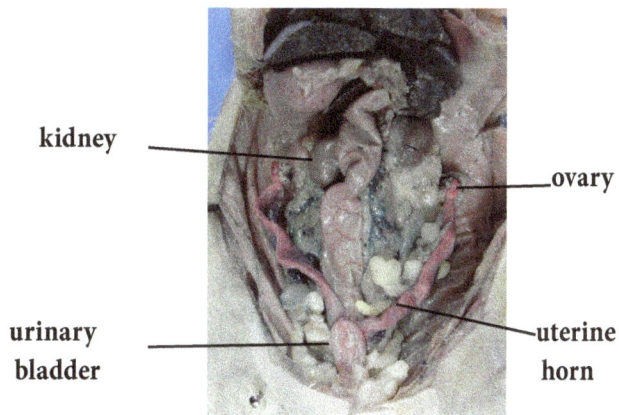

FIGURE A2.19. Internal anatomy of the female rat.

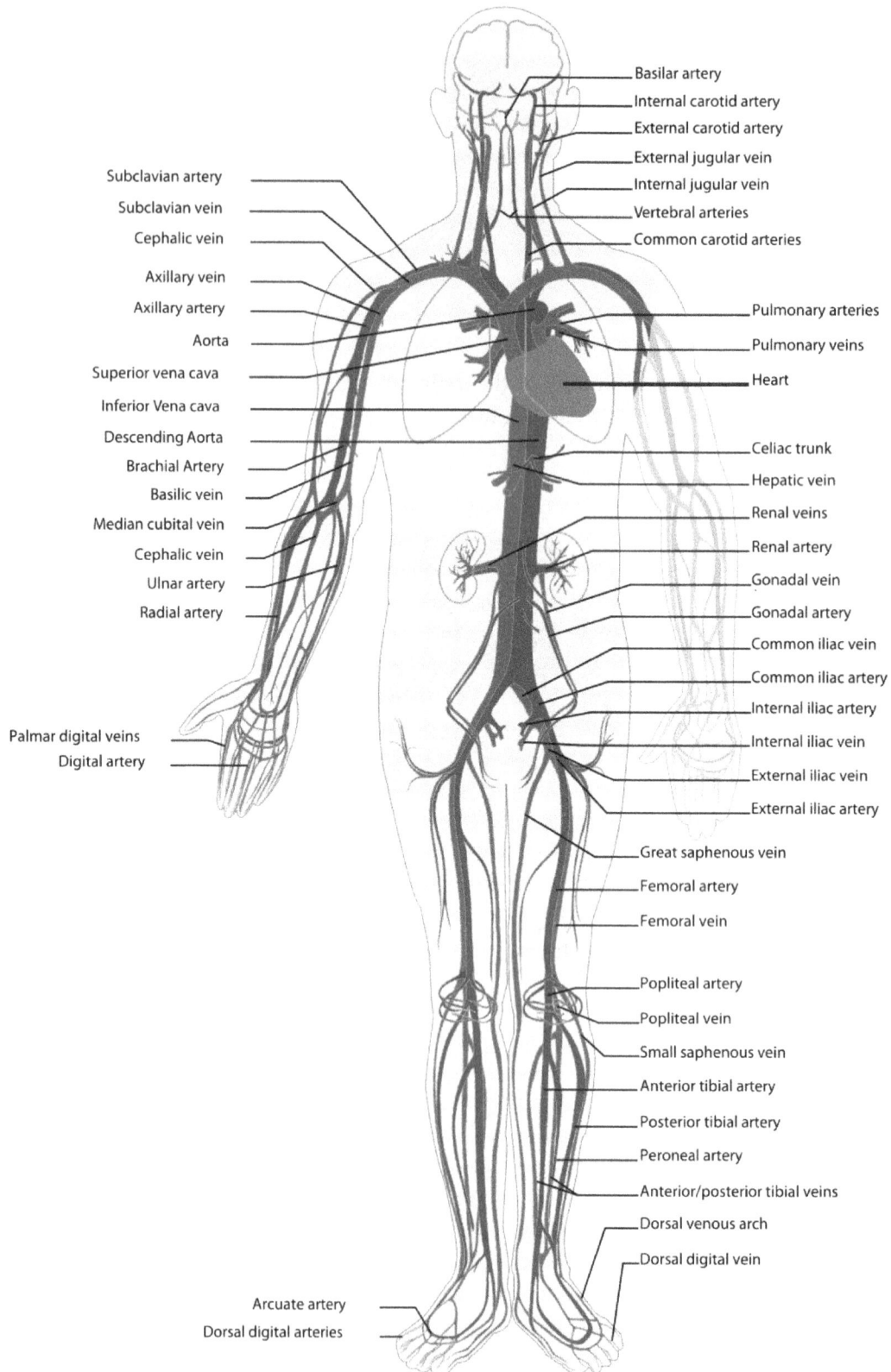

Basilar artery
Internal carotid artery
External carotid artery
External jugular vein
Internal jugular vein
Vertebral arteries
Common carotid arteries

Subclavian artery
Subclavian vein
Cephalic vein
Axillary vein
Axillary artery
Aorta
Superior vena cava
Inferior Vena cava
Descending Aorta
Brachial Artery
Basilic vein
Median cubital vein
Cephalic vein
Ulnar artery
Radial artery

Pulmonary arteries
Pulmonary veins
Heart

Celiac trunk
Hepatic vein
Renal veins
Renal artery
Gonadal vein
Gonadal artery
Common iliac vein
Common iliac artery
Internal iliac artery
Internal iliac vein
External iliac vein
External iliac artery

Palmar digital veins
Digital artery

Great saphenous vein
Femoral artery
Femoral vein

Popliteal artery
Popliteal vein
Small saphenous vein
Anterior tibial artery
Posterior tibial artery
Peroneal artery
Anterior/posterior tibial veins
Dorsal venous arch
Dorsal digital vein

Arcuate artery
Dorsal digital arteries

FIGURE A2.20. **The human circulatory system.**

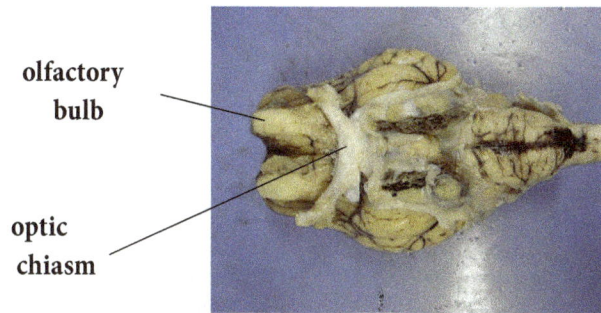

FIGURE A2.21. Sheep brain, ventral view.

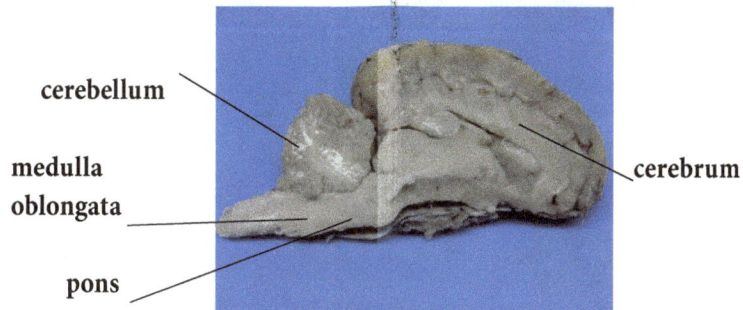

FIGURE A2.22. Sheep brain, left sagittal view.

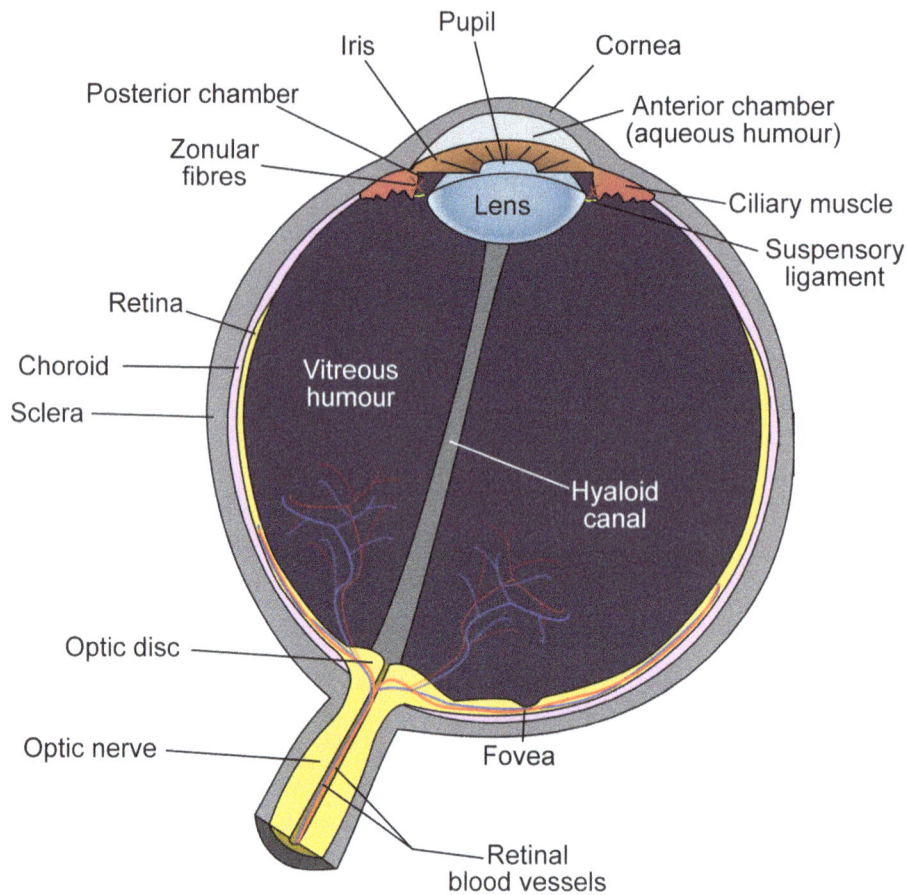

FIGURE A2.23. **Diagram of the human eye.**

FIGURE CREDITS

Figure A.2.3: Source: http://commons.wikimedia.org/wiki/File:Annelida_Oligochaeta_%28Lumbricus%29_Earthworm.jpg.

Figure A.2.20: Source: http://commons.wikimedia.org/wiki/File:Circulatory_System_en.svg.

Figure A.2.23: Source: http://commons.wikimedia.org/wiki/File:Schematic_diagram_of_the_human_eye_en.svg.

www.ingramcontent.com/pod-product-compliance
Lightning Source LLC
Chambersburg PA
CBHW082035230326
41598CB00081B/6513